AF904488

*Older People in Natural Disasters*

# *Older People in Natural Disasters*

## *The Great Hanshin Earthquake of 1995*

Junko Otani

Kyoto University Press

TRANS PACIFIC PRESS

The publication of this book was made possible by a Grant-in-Aid for Publication of Scientific Research Results from the Japan Society for the Promotion of Science.

First published in 2010 jointly by:

Kyoto University Press
Kyodai Kaikan
15-9 Yoshida Kawara-cho
Sakyo-ku, Kyoto 606-8305, Japan
Telephone: +81-75-761-6182
Fax: +81-75-761-6190
Email: sales@kyoto-up.or.jp
Web: http://www.kyoto-up.or.jp

Trans Pacific Press
PO Box 164, Balwyn North, Melbourne
Victoria 3104, Australia
Telephone: +61-3-9859-1112
Fax: +61-3-9859-4110
Email: tpp.mail@gmail.com
Web: http://www.transpacificpress.com

Copyright © Kyoto University Press and Trans Pacific Press 2010.

Edited by Ashley Davis.

Set by Digital Environs, Melbourne.

Printed in Melbourne by BPA Print Group.

Cover photo: AP/AFLO.

**Distributors**

**Australia and New Zealand**
DA Information Services/Central Book Services
648 Whitehorse Road
Mitcham, Victoria 3132
Australia
Telephone: +61-3-9210-7777
Fax: + 61-3-9210-7788
Email: books@dadirect.com
Web: www.dadirect.com

**USA and Canada**
International Specialized Book Services (ISBS)
920 NE 58th Avenue, Suite 300
Portland, Oregon 97213-3786
USA
Telephone: 1-800-944-6190
Fax: 1-503-280-8832
Email: orders@isbs.com
Web: http://www.isbs.com

**Asia and the Pacific**
Kinokuniya Company Ltd.

*Head office*:
38-1 Sakuragaoka 5-chome
Setagaya-ku, Tokyo 156-8691
Japan
Telephone: +81-3-3439-0161
Fax: +81-3-3439-0839
Email: bkimp@kinokuniya.co.jp
Web: www.kinokuniya.co.jp

*Asia-Pacific office*:
Kinokuniya Book Stores of Singapore Pte., Ltd.
391B Orchard Road #13-06/07/08
Ngee Ann City Tower B
Singapore 238874
Telephone: +65-6276-5558
Fax: +65-6276-5570
Email: SSO@kinokuniya.co.jp

All rights reserved. No production of any part of this book may take place without the written permission of Kyoto University Press or Trans Pacific Press.

ISBN 978-1-920901-89-9

# *Contents*

# *Figures*

# *Photographs*

# *Tables*

# *Abbreviations and Acronyms*

| | |
|---|---|
| ABC | Asahi Broadcasting Corporation |
| GH | General Housing |
| IMR | Infant Mortality Rate |
| KAST | Kurihama alcoholism screening test |
| LSA | Life Support Adviser for elderly people |
| MBS | Mainichi Broadcasting |
| MHLW | Ministry of Health, Labour and Welfare |
| MOC | Ministry of Construction |
| MOHW | Ministry of Health and Welfare |
| MOHLW | Ministry of Health, Labour and Welfare |
| MOL | Ministry of Labour |
| NGO | Non-Governmental Organisation |
| NHK | Nihon Hōsō Kyōkai (Japan Broadcasting Corporation) |
| PRH | Public Reconstruction Housing |
| QSR NVivo | Qualitative Solutions & Research NUD*IST (Non-numerical Unstructured Data * Indexing, Searching and Theorizing) Vivo |
| TFR | Total Fertility Rate |
| TSH | Temporary Shelter Housing |

# *Glossary*

| | |
|---|---|
| *Dokkyo rojin* | Old person living alone |
| *Fukkō jūtaku* | Public Reconstruction Housing (PRH) |
| *Fureai* | Human contact, relationship |
| *Gyōsei* | Public administration, government |
| *Ie seido* | Households system |
| *Ikigai* | Purpose and meaning in life |
| *Jichikai* | Self-governing body |
| *Jichikaichō* | Chief of self-governing body |
| *Kaigo Hoken* | The Social Care Insurance Scheme (SCIS) |
| *Kasetsu* | Temporary Shelter Housing (TSH) |
| *Kodokushi* | Isolated death, dying alone |
| *Kokoro no kea* | Mental health |
| *Minsei iin* | Unpaid welfare commissioner |
| *Nagaya* | One story hutted apartment comprised of several small flats |
| *Renkei* | Coordinated collaboration |
| *Sawa-kai* | Tea party |
| *Seikatsu fukkō* | Life reconstruction |
| *Seikatsu fukkō shien in* | Life reconstruction support adviser |
| *Shakai fukushi hojin* | Social Welfare Corporation |
| *Tetsu no tobira* | Iron door |
| *Tokuyō (Tokubetsu Yōgo Rōjin Home)* | Special home for poor eldelry people |
| *Tsunagari* | Being related to someone |
| *Yakuin* | Committee member |
| *Yūai homon* | 'Friendship and love' home visit |

# *Preface*

This book grew out of a concern for the ageing society and economic stagnation experienced in Japan. It reports social scientific research on elderly people in urban areas who are poor and have no functioning family, a group that will be of increasing future concern in Japan and many other countries. The study focuses on elderly people who lost their homes in the 1995 Great Hanshin Earthquake and were repeatedly relocated to various types of housing schemes in the following years. It follows the long-term process of recovery for these people by looking at the development of aged communities in temporary shelter housing (TSH) created in Kobe after the earthquake. It also observes the relocation of some of the poorest and most vulnerable of these people into state-subsidized public reconstruction housing (PRH), with reference to individuals and groups involved in housing and community work.

The research involved media analysis, secondary analysis of the Hyogo Health Survey, and ethnographic research at selected TSH and PRH compounds in central and suburban Kobe. Analysis drew on qualitative and quantitative approaches and the research was the first to utilize NVivo computer assisted qualitative analysis software in a Japanese research setting. Discourse analysis of media data was an important component of this study of Japanese society.

The secondary analysis of data from annual Hyogo surveys illustrated the way that the surveys changed over time in relation to changing social contexts. The media analysis identified the main foci of public attention and how this also changed over time. Elderly people, especially those who lived alone, received considerable attention in the media. Along with original observational research conducted in three sample sites, this research helped to question received knowledge about post-earthquake reconstruction and the effectiveness of TSH and PRH as solutions to the problem of housing earthquake survivors. Case studies of the media's presentation of loneliness and *kodukushi* showed how phenomena that are perceived to be new are often built up to become important aspects of culture. The research revealed that gender perspectives were lacking in public surveys, despite being crucial to my qualitative analysis of media data and field sites.

Disasters are long drawn out events for vulnerable elderly people, especially those without money or family connections. Official statistics and the media contribute to the received knowledge about the recovery process, while workers on the ground reproduce many of these views and add some prejudices of their own. In its conclusion, this book considers the policy implications of this relationship.

So many people have played an important part in this research and it is not possible to acknowledge everyone by name. However, I should like to express my special gratitude to some of those who have assisted me:

Foremost, the Kobe people, especially the residents of TSH and PRH as well as the many volunteers, public health nurses and other officials. Special thanks go to the people I met in the field who's names I refrain from listing to protect their anonymity.

The Salzburg Seminar for the scholarship to participate in the Aging Seminar, where I met Mr Shiro Yamazaki, then Director of Planning Division of Welfare for the Elderly, MOHW. Mr Yamakazi not only provided me with a lot of information but introduced me to many key informants.

My supervisor Dr. Gail Wilson for providing competent academic advice with solid knowledge, inspiration and encouragement that pushed me to keep going. Her support continually pulled me up to a higher level each time that I said I could not do any better.

Dr. Silvana di Gregorio, Prof. Lyn Richards and Dr. Pat Bazeley for their guidance and actual exercises that helped me to utilize NVivo in the analysis of my qualitative data. The comments they offered on my original manuscript lead to many new surprises and discoveries.

The joint Japan and World Bank Graduate Scholarship for funding the first two years of my study. I also thank the United Nations University Institute of Advanced Studies (UNU/IAS) for their part in the writing-up stage.

Dr. Janet Hohnen of the World Bank for her support and for giving me the opportunity to pursue a PhD in London. It was a wonderful break to go with her as a consultant on World Bank missions and health projects in the Solomon Islands and China after I left the World Bank for study in London. I am also grateful to the World Bank for the financial support when I returned to student life.

JOCS for funding my local travel during fieldwork. I am also grateful to JOCS workers for their emotional support, especially Ms Kiyoko Okawa, public health nurse, who in 1999 went to work with the earthquake-affected community in Taiwan after our work in Kobe.

The World Bank for helping me to develop my review of the literature on population aging and the World Health Organization China Representative Office of the Western Pacific Region for giving me the opportunity to work full-time in the area of ageing and disaster relief alongside the International Federation of the Red Cross and the World Food Programme.

My many friends and fellow PhD students in various colleges of the University of London, London School of Economics and Political Science, London School of Hygiene and Tropical Medicine, University College London, Institute of Education and other universities and colleges who gave valuable support and guidance. Special thanks go to Dr. Sukohntha Kongsin (Jieb), Dr. Sandra Reyes-Frausto (Beaman), Dr. Zsuzsa Varvasovszky, Dr. Bingqin Li and Dr. Hyun Bang Shin for their friendships as we all went through PhD process together.

The London School of Economics and Political Science, headed by Prof. Anthony Giddens, for making me feel at home in London, even when I was working full time in China. My experiences there enabled me to see Japan not only through Asian and American eyes, but through the eyes of Europe as well.

The Japan Society for the Promotion of Science (JSPS), who made this publication possible. The editors at Kyoto University Press and Trans Pacific Press were of great help in preparing the manuscript. I am grateful to the wonderful and careful editing, although remaining errors are my fault. JSPS also funded the Japanese edition (published by Kyushu University Press, 2006) and in Chinese edition (to be published by SMC Publishing, Taiwan, 2010). JSPS also financed my presentation of part of this research at the Sixteenth International Sociological Association (ISA) World Congress of Sociology in Durban, South Africa, July 2006.

Once again, Dr. Gail Wilson, Reader Emeritus in Social Policy and Ageing, LSE. Without her support and guidance, the writing process would not have been possible, not only in practical terms but also in terms of stimulating my thinking and keeping up my motivation. As a woman she was a role model who was always there for me despite being very busy. She would intervene at the right time but would otherwise make me stand on my own; I can therefore say this is my thesis and that I learnt a lot through writing it. Although she did not always need to show it, when I was struggling she was wise, kind-hearted and knowledgeable. As a result of my relationship with her, I have become more mature and responsible. I will miss being her student but I hope this is the beginning of our friendship.

Finally, my family for their patience and for giving me all that is necessary to keep going.

# *Chapter One*
# *Introduction*

> 'The public reconstruction of housing after disasters is especially associated with the care of highly aged social groups. But it will soon be an issue all over Japan, not only in the disaster-affected area.'
>
> Life Support Advisor on visiting barrier-free housing, Kobe (1999)

This study addresses two global trends occurring concurrently: urbanization and population ageing. Its focus is on the community provision of housing and health care for elderly people in contemporary Japan. It compares two kinds of dwellings built following the Great Hanshin Earthquake that shook the Kobe and Osaka area in 1995. The isolation of elderly people after the earthquake reflected a much deeper problem in the provision of care for elderly people in Japan (Iliffe, 1997: 519). It has since been seen as a problem for the whole of Japan.

This book is concerned with the situation of elderly people living in urban areas who are poor and have no functioning family. This particularly vulnerable group will be of increasing concern in Japan and many other countries (Wilson, 2000: 117). The book focuses on those elderly people who lost their homes in the Great Hanshin Earthquake of 1995 and were repeatedly relocated to various types of housing schemes in the following years. It analyses the post-earthquake reconstruction process by looking at media coverage in TV programs and broadsheet newspapers, reexamining the Hyogo Health Surveys, and reporting on ethnographic research performed in selected temporary shelters and public housing compounds located in central and suburban areas. It makes special reference to the efforts that were made to provide housing and community support for elderly people.

This first chapter presents the background of the study, focusing on population aging in the context of the Japanese health and welfare system and community care. It demonstrates how the support and care of elderly people has received increased attention in recent years. Following this, it specifies the research questions and outlines the structure of the rest of the book.

# Background

### *Population ageing in Japan*

Since the early 1990s, rapid population ageing in Japan has been accompanied by economic stagnation. The 1997 census revealed for the first time that there were more people aged 65 and older than aged 15 and younger.[1] Since the 1970s, the issues relating to population ageing have received increasing government and public attention (Campbell, 1992: 139). A number of policies and programs for the elderly have been implemented, but still much more needs to be done.

Japan has the most rapidly ageing population in the world. The trend towards ageing started later but has progressed faster than in any other country. Between 1948 and 1995, life expectancy at birth improved from 56 to 76 years for Japanese males and from 59 to 83 years for females (Ogawa and Retherford, 1997: 62). By 2001, it was 77.9 years for males and 84.7 years for females (WHO, 2002). Mortality has dropped for all ages. The Infant Mortality Rate (IMR) has fallen particularly quickly. The Total Fertility Rate (TFR) has also dropped from 3.6 in 1950 to 1.4 in 1995, far below the replacement level of 2.1 (Japan MOHW, 1997).

The proportion of those aged over 65 rose from 7% in 1970 to 14% in 1995. While it took France 125 years to experience comparable demographic change, in Japan it occurred over just 25 years (Japan MOHW, 1997). Japan was the first country to experience population aging within the space of three generations (Ochiai, 1997[2]: 65). Now, the proportion of people aged over 75 is increasing even more rapidly than those aged between 65 and 75. It is projected that similar rapid population ageing will occur in other Asian countries such as China, India, Thailand and Sri Lanka (Phillips, 1992).

Tables 1.1 and 1.2 show the projected population trends between 1990 and 2000. They demonstrate that population ageing was found to be more rapid in 2000 than in 1990. The most significant reductions were observed in the 0–14 age group, whereas little change was evident in the older age groups. Elderly people represented a much greater proportion of the overall population in 2000 than they did in 1990.

Japan's situation is unique when compared to other Western countries because it was the first non-Western country to experience rapid industrialization. The experience of population aging has affected Japan's social, cultural and political institutions in ways that might be different to other countries.

The proportion of Japanese who lived in cities increased from 28% in 1945 to 77% in 1985 (Sonoda, 1988). By 2002, it had increased to 79%

Table 1-1: Japanese population projections for four age groups (in thousands of people), 1990

| | 0–14 | 15–64 | 65–74 | 75 and over | Total | Average age |
|---|---|---|---|---|---|---|
| 1970 | 25,153 | 72,119 | 7,393 | 2,237 | 104,665 | 30.3 |
| | (24.0%) | (68.9%) | (7.1%) | (2.1%) | (100%) | |
| 1995 | 22,387 | 87,168 | 18,009 | 6,986 | 127,565 | 38.8 |
| | (17.5%) | (68.3%) | (14.1%) | (5.5%) | (100%) | |
| 2000 | 23,591 | 86,263 | 21,338 | 8,452 | 131,192 | 39.8 |
| | (18.0%) | (65.8%) | (16.3%) | (6.4%) | (100%) | |
| 2025 | 22,075 | 81,102 | 31,465 | 17,367 | 134,642 | 43.3 |
| | (16.4%) | (60.2%) | (23.4%) | (12.9%) | (100%) | |

Note: Figures are as of 1 October for each year.

Source: Adapted from the 1985 national census and population estimates, Statistics Bureau, Management and Coordination Agency (Bentelspacher, 1994), and the 1990 future population estimates for Japan, Institute for Population Problems (Kinoshita, 1992).

Table 1-2: Japanese population projections for three age groups (thousands of people), 2000–2050 (estimated in 2002)

| | 0–14 | 15–64 | 65 and over | Total |
|---|---|---|---|---|
| 2000 | 18,505 | 86,380 | 22,041 | 126,929 |
| | (14.6%) | (68.1%) | (17.4%) | (100%) |
| 2025 | 14,085 | 81,102 | 31,465 | 134,642 |
| | (11.6%) | (60.2%) | (23.4%) | (100%) |
| 2050 | 10,842 | 53,889 | 35,863 | 100,593 |
| | (10.8%) | (53.6%) | (35.7%) | (100%) |

Source: National Institute of Population and Social Security Research (2002).

with an annual growth rate of 0.3% (UN ESCAP, 2002). Such urbanization has implications for social security and traditional family structures as occupational opportunities in cities are not the same as in rural areas and housing for urban migrants tends to be smaller and is usually inappropriate for multi-generational households.

### *Economic trends*

Japan is in the throes of a ten-year recession, the longest and most severe in its post-war economic history. The common Japanese practice of full employment for life has become difficult to sustain, and early retirement has been a trend for more than a decade. The numbers of unemployed and homeless people are increasing (Phillips, 1992).

The unemployment rate in May 1998 was 4.2%, the highest since the end of World War II. By December 2001 it had peaked at 5.5%, dropping to 5.2% in April 2002 (OECD, 2002). Because of the recession, joblessness is more severe for marginalized older people in the labor force. Workers between the age of 45 and 54 are the worst affected by unemployment; in May 1998 8.4% were unemployed, compared to 7.1% in 1997.

The experience of unemployment for older people has been made more severe by the reduction of the budget for old age welfare, a policy change linked to the recession. This might help to explain the substantial increase in suicide rates among men in their 40s and 50s, increasing in 1998 by 50% from 20,000 annual cases to over 30,000. In 1998, *Asashi Shinbun* newspaper reported that a main cause of depression was redundancy after company restructuring. As Table 1.3 shows, as the number of old people increase, the total amount and proportion of government welfare spent on the elderly also increases.

**Income of elderly individuals**

According to the MOHW (Ministry of Health and Welfare[3]) (1997), elderly Japanese people are wealthier overall than younger generations. They enjoyed the fastest income gains of any age group in recent years, primarily because of expansions in pensions that began in the mid-1970s. Between 1981 and 1996 the proportion of people aged 60 and over who depended on their pension as their primary income source increased from 35% to 57% and the proportion who reported that work was their primary income source declined from 31 to 22%[4] (Ogawa, 1997). The pre-recession expansion of pension systems resulted in fewer elderly Japanese people reporting economic difficulty than in any other industrialized counties (Palmore, 1999). Although there is little gender-sensitive data on pension distribution in Japan[5], Table 1.4 shows that women did very badly in terms of public pensions in the 1990s. It is unlikely that their position has greatly improved.

Table 1.5 shows that public pensions are the largest source of income for elderly people, with private pensions making up only a small component. Surprisingly, in view of Japan's relatively high saving rate (Ogawa, 1997), the table also shows that the proportion of elderly people who mention savings as a source of income is no higher in Japan than in the US or Germany. This may be because the savings of elderly Japanese are tied up in home ownership to a greater extent than in the other two countries (Ogawa, 1997: 73). Or it might be because people are less prepared to report savings.

Table 1-3: Changes in the provision of welfare to the elderly in Japan, 1973–1998

| | Expenditure on elderly people's welfare (100 million yen) | Total welfare expenditure (100 million yen) | Proportion of welfare for old people (%) |
|---|---|---|---|
| 1973 (Showa 48) | 15,641 | 62,587 | 24.99 |
| 1985 (Showa 60) | 188,287 | 356,798 | 52.77 |
| 1995 (Heisei 7) | 407,109 | 647,314 | 62.89 |
| 1998 (Heisei 10) | 477,865 | 721,411 | 66.24 |

Source: National Institute of Population and Social Security Research (2001).

Table 1-4: Monthly public pension amount by sex

| | Men | Women |
|---|---|---|
| Less than 100,000 yen | 3.1% | 50.5% |
| 100,000–150,000 yen | 29.8% | 39.5% |
| 150,000–200,000 yen | 39.1% | 9.1% |
| 200,000–250,000 yen | 25.5% | 0.9% |
| More than 250,000 yen | 2.5% | 0.0% |
| Total | 100.0% | 100.0% |
| Average monthly amount | 172,000 yen | 107,000 yen |

Source: Fujii (1993: 192).

It is known that savings, at least 21% of the income of elderly Japanese, comprise a higher proportion of income in Japan than in any other OECD country. The average saving rate in Japan between 1984 and 1993 was 32.8%, whereas in the UK it was only 15.9%. This is especially remarkable given that Japan's saving rate peaked around 1970 and thereafter declined quickly until 1984 (Hayashi, 1992: 63–78). This decline is most obvious among younger generations.

Work comprised 35% of elderly Japanese household income in 1996. Table 1.6 shows that the rate of participation in the labor force for elderly Japanese men was high by international standards. But the employment rate of elderly people is falling, even though the desire to work late in life is rising (MOHW, 1997, Bass and Oka, 1995).

Table 1-5: Income sources for people aged 60 and over by country, 1996

| | All nominated sources by percentage | | | Main nominated source only by percentage | | |
|---|---|---|---|---|---|---|
| **Income source** | **Japan** | **US** | **Germany** | **Japan** | **US** | **Germany** |
| Work | 35 | 26 | 7 | 22 | 16 | 5 |
| Public pensions | 84 | 83 | 84 | 57 | 56 | 77 |
| Private pensions | 8 | 33 | 24 | 2 | 13 | 10 |
| Savings | 21 | 24 | 21 | 2 | 2 | 2 |
| Assets | 11 | 34 | 12 | 3 | 9 | 2 |
| Children | 15 | 3 | 3 | 4 | 0 | 0 |
| Public assistance | 1 | 2 | 1 | 0 | 0 | 1 |
| Other | 4 | 7 | 4 | 2 | 2 | 2 |
| No answer | 0 | 2 | 0 | 8 | 4 | 2 |

Note: Results are based on self-reports. General income is greater than 100% (first set of columns) because respondents often indicated more than one source. The last set of columns reports only the main source of income so each column equals 100% within a rounding error. The distinction between savings and assets is not clear-cut; Japanese view savings as money in a savings account.

Source: Management and coordination agency (in Ogawa and Retherford, 1997: 73).

Table 1-6: Comparison by selected countries of the proportion of elderly people who work with the proportion who desire to work, 1994

| | Japan | USA | Germany | Korea | Thailand |
|---|---|---|---|---|---|
| Proportion of the elderly who work | 43.6% | 23.5% | 6.8% | 33.6% | 37.1% |
| Age: proportion of age-group who work | | | | | |
| 60–64 | 63.9 | 44.3 | 17.1 | 56.5 | 54.9 |
| 65–69 | 44.8 | 29.4 | 7.5 | 42.6 | 39.3 |
| 70–74 | 38.3 | 18.7 | 1.1 | 26.9 | 23.4 |
| 75–79 | 20.6 | 7.7 | 0.9 | 19.9 | 19.2 |
| 80 + | 13.7 | 2.2 | 1.3 | 10.1 | 11.6 |
| Desire to work | 89.4% | 90.0% | 69.4% | 79.9% | 82.8% |

Source: Japanese MOHW (1997: 51).

Table 1-7: Labor force participation rates of people aged over 65 years in selected countries, 1985

| | Males | Females |
|---|---|---|
| Japan | 37.0 | 15.5 |
| Germany | 5.1 | 2.1 |
| UK | 8.2 | 3.0 |
| USA | 15.2 | 6.8 |

Source: Labor force statistics, OECD (Hagemann, 1990).

Job loss for men often results in loss of *ikigai* (value of life[6]) and opportunities to develop social networks. People think they need to continue to work in a daily routine in order to maintain their health. Indeed, evidence suggests that those who work longer enjoy better health in their old age (OECD, 1996). But it should be noted that those who are healthier are often able to work later in life. The stigma of not working is particularly severe in Japan, especially among the generation that saw post-war economic growth. It is common for men to put loyalty to their company before their family life, a practice called *kaishashugi*, or company-firstism (Harada, 1996: 6). Company loyalty was supported by the Japanese practice of life-long employment, but permanent employment might now be less common, especially as the economy is no longer growing continuously (Matsuzaka, 2002:1).

The gender difference in labor force participation is significant (see Table 1.7). Although it is increasing, the post-war participation of women in the labor force has been low, affecting not only women's access to wealth and income but also on benefits such as health insurance. Therefore, there are significant gender differences in terms of income, living arrangements, and access to welfare services (Pennec et al., 1996). Table 1.8 shows that more elderly women than men depended on a child as their primary source of income in 1988. Furthermore, in 1997 *Asashi* newspaper reported that the pension entitlement was unequal between genders. In 1997 the average monthly pension for a man (JPY 207,000) was almost double that of a woman (JPY 119,000).

Table 1.9 shows that the likelihood that an elderly single women living alone will fall into poverty is much higher in Japan than in the UK, USA and Germany. It is much more difficult for women to purchase a flat or to borrow the money to build a house. Pension entitlements are not decided by a married woman's household contribution, but by her husband's (Fujii,

Table 1-8: Proportion of people aged 65–69 years dependent on a child's income, 1980 and 1988

| | Men with work | Men without work | Women with work | Women without work |
|---|---|---|---|---|
| 1980 | 14 | 37 | 37 | 52 |
| 1988 | 13 | 23 | 28 | 34 |

Source: Miyajima (1992).

Table 1-9: Proportions of women in the lowest income quintile, 2000

| | Age 18 and above | | Age 75 and above | | |
|---|---|---|---|---|---|
| | Entire population (M&W) | All older old (M&W) | Single women living alone | Women living with spouse only | Single persons living with others (and the person is not household head) |
| Germany | 20 | 31 | 43 | 15 | ** |
| Japan | 20 | 34 | 79 | 59 | 18 |
| UK | 20 | 39 | 47 | 39 | 13 |
| USA | 20 | 35 | 56 | 23 | 19 |

** = data not available.

Sources: OECD calculations based mainly on data from the Luxembourg income study (2002: 49).

1993). Also, the likelihood of being widowed is much higher for Japanese women than for men; elderly women of this generation tended to marry men ten years their senior. In effect, the average length of widowhood of Japanese women is about twenty years. Widowhood is a factor in old age poverty (Barusch, 1994).

As well as gender inequalities, the MOHW White Paper (1997) also reported a large income differential between elderly households. The ratio of low-income elderly households was high compared to all households, suggesting that there was a large gap between the haves and the have-nots.

### *Family trends and changes*

Population ageing and urbanization have implications for family structures that are intertwined with social security systems (Ogawa, 1989, Boling, 1998). Changes in household occupations and, subsequently, in income patterns, have transformed what Japanese people expect from the systems that support elderly people (Hashimoto, 1996). Expectations about independence can vary in important ways between generations. Traditionally, under the *ie* family system[7] the eldest son would inherit the family property and, in return, he and his family would live with his parents and look after them (Hiroshima, 1992, Otani, 1998, Koyama 1996). Although *ie* was formally abolished during Japan's post-war democratization (Maykovich, 1979:381–410), the practice remains (Yamamoto and Wallhagen, 1998) but has weakened considerably over time (Ogawa and Retherford, 1997, Long and Harris, 2001, Kano, 2000).

It is traditional for elderly people to be seen as independent and not a burden on society, therefore stigmatizing those who need support (Powell and Anesaki, 1990). Some elderly people wish to be independent so as not to be a burden on their children, or to maintain their own freedom (Hashimoto, 1996: 94–98), hoping only to count on their children when they become very frail (Wilson, 2000: 89). Some parents also fear that their children might burn out from taking care of them before they really need to do so (Hashimoto, 1996). When elderly people need daily assistance, deciding on what assistance to take is complex. Despite recent trends to promote formal care, cultural barriers are still apparent; for example, an elderly person might feel ashamed to receive formal care if they have a daughter-in-law (Yamamoto and Wallhagen, 1997). The role of family care is largely taken by women. In 2000, a total of 72.7% of caregivers were women, and one-third were themselves aged 65 and above. Wives (19.8%), daughters (19.7%) and daughter-in-laws (29.0%) were reported to be the most common caregivers, while husbands (8.4%) and sons (9.5%) were relatively uncommon (Fukushi channel, 2002).

The rate of co-residence in Japan is falling, although it is still much higher than in any other developed country, In 1960, as many as 60% of elderly people lived in households with other generations (MOHW, 1997). Table 1.10 shows the percentage spread of household types that included members over 65 years of age in 1988 (Sonoda, 1988). Since then, the proportion of one-person households has increased, along with a decrease in the proportion of three-generation households.

The proportion of elderly Japanese parents living with married children is exceptionally high compared to other industrialized countries and it is decreasing at a rate of less than 1% annually (Naoi, 1990). The practice of co-habitation[8] seems to be a major safety net, but it might also be a major reason for the slow development of public programs for the assistance of elderly people. Likewise, housing to meet the needs of elderly people has not been well developed (Izuhara, 2000a: 89). However, this might have less

Table 1-10: Proportion of households by type that include members over 65 years of age, 1975–2005

| Years | One-person households | Married couples only | Two-parent households and single-parent households | Households consisting of three generations | Other households | Total |
|---|---|---|---|---|---|---|
| 1975 | 9 | 13 | 10 | 54 | 14 | 100 |
| 1985 | 12 | 19 | 11 | 46 | 12 | 100 |
| 1995 | 17 | 24 | 13 | 33 | 12 | 100 |
| 2005 | 22 | 29 | 16 | 21 | 11 | 100 |

Source: MHLW (2005).

to do with a lack of effective government, and more a problem stemming from cultural values. Some older people, while being eligible for public assistance, refuse to accept it, preferring instead that a family member, especially the wife of the eldest son, provide daily assistance (Yamamoto and Wallhagen, 1998).The alternative is to risk the loss of face in front of their extended family and neighbors. However, this does not necessarily mean that parents-in-law are emotionally close to their son's wife (Yamamoto and Wallhagen, 1998).

The spread of the nuclear family in Japan has resulted in a decrease in the percentage of households co-habited by elderly parents and their children. 40% of elderly people now live either alone or as a couple (see Table 1.10). It is becoming increasingly difficult to expect children and extended family to support the elderly (MOHW, 1997, Maeda, 1989). Although the number of children living with their parents is declining, about half choose to live less than an hour away. It is now common for the younger generation to consider the question of living with and supporting their parents as separate issues (MOHW, 1997). Co-residence with elderly parents is much lower in big cities compared to rural areas[9].

In conclusion, Japan is a rapidly aging society that now faces a range of challenges in providing for its elderly population. Individual savings have fallen and family support is weaker than in the past. As a result of industrialization and urbanization, employment opportunities have changed and elderly people are increasingly dependent upon pension benefits and state support. This dependency was exposed further by the Great Hanshin Earthquake, the victims of which were disproportionately women and disproportionately poor. The next section will briefly consider how the Japanese welfare system has faced this growing challenge.

### *Social security in Japan*

This section describes the evolution of the Japanese social security system since the 1970s. According to national policy, the health and welfare of the elderly must be secured, especially since the current generation suffered during war time and worked hard to reconstruct the nation. According to Campbell, Japan has been considerably more active than most other developed countries, as well as other countries in Asia, in securing the employment of the elderly, providing for recreational programs and in aiding families living with dependent partners. However, it was substantially behind in such areas as social services and housing policy (Campbell, 1992: 18). This fact is receiving increased attention as Japan goes into the next stage of the welfare state to ensure the quality of life in old age.

#### The design of the Japanese social security system

In 1926, Japan became the first non-Western country to introduce social insurance. The Japanese social insurance system has two tiers (MOHW, 1996): German-styled workplace-based insurance and British-styled community-based insurance. Work-based insurance is based on the occupation of the insured person, while community-based insurance is based on the geographical location of the insured person's residence. These two tiers each comprise a number of different insurer groups. Work place insurance is the older form in Japan (Ikegami and Hasegawa, 1995), but there has been a gradual shift to the latter universal tax-based system (Hiroi, 1998).

There is a widening gap between those covered by the former system and those covered by the latter system. Although people covered by the latter system make smaller contributions, they generally receive a lower standard of pension and medical cost coverage. Those who cannot afford insurance, including the self-employed, unemployed and a high proportion of elderly people, are covered by the community-based NHI (National Health Insurance) scheme (Hoshino, 1996). The NHI system receives no employer funding and has a high proportion of low-income subscribers. The system is not self-funding, but relies on a large state subsidy. The benefits are fixed nationally and contributions are centralized and then distributed at lower levels. The government subsidizes this largely because of a shortfall in contributions from the population groups.

According to official sources Japan has had universal social security since 1961 (Hoshino, 1996, Yamasaki, 1998). However, it must be noted that the whole population is covered only in theory. There are people who are not covered at all, and some who are covered do not receive enough secure the minimum standard of living.

**Policy responses to population ageing**

Since population ageing was first noted in the 1970s, there has been a sharp change in the direction of public policies in Japan. Partly because of increased media scrutiny, the public grew very concerned about demographic change. As a result, health care expenditure for the elderly increased with the introduction of free medical care and the expansion of pension schemes in 1973.[10] Co-payments replaced free medical care in 1985. Co-payments are small, especially for elderly people, but they are increasing. They are also higher for those covered by NHI, therefore placing a greater burden on women and low-income groups.

In 1992, to cover the effects of the bursting of the economic bubble in 1990, the central government imposed a consumption tax of 3% to finance the NHI scheme. This was raised to 5% in 1997.

**Resource planning in response to population aging**

The 1989 Gold Plan was designed to meet the resource demands of an ageing society. The plan was intended not only to improve the quality of welfare, but also to save medical expenditure in the longer term. The aim was to increase the number of nursing home beds available to elderly people in need of short-stay care and institutionalization, as well as to increase the number of home help services for elderly people. Although this was to be achieved within ten years, public expenditure increased dramatically and the targets had almost been reached within five years and revised targets were set in 1994.

In Japan, residential care homes are widely seen as unacceptable but hospital care is not (Wilson, 2000: 146). Subsequently, the practice of *shakaiteki nyuin* (social hospitalization)[11], where elderly people are placed into hospital care rather than nursing homes, is common, increasing the expenditure of government on the hospital system (Ikegami and Campbell, 1996). Hospital stays are affordable with little, if any, charges for medical services, and hospitalization can then be presented not as disrespect to the elderly family member, but as a medical necessity (Campbell and Ikegami, 1998). The Japanese government's goal is to abolish this practice (Campbell and Ikegami, 1998, Yamamoto and Wallhagen, 1997 and 1998). A major controversy is how to abolish it and what alternatives should be provided.

Since the introduction of free medical care for elderly people and the expansion of the pension that occurred between the early 1970s and 1985, many younger people feel less the burden of taking care of their parents (Yazawa and Kunihiro, 1999). Government efforts to shift some of the burden of caring for the elderly back to families in order to reduce national spending have met with public resistance, especially from women's groups

(Campbell, 1992: 221). Despite this opposition, the government has tried to utilize the sense of cultural obligation and community co-operation often felt by families to control the growing costs associated with Japan's aging population (Ogawa, 1999).

### *Community care in Japan*

When government says that the elderly are cared for by their families, it generally means women relatives (Osawa, 1995: 135–136). But alongside this expectation is *minsei iin*, the volunteer system of community carers, many of whom are men. In its long history, *minsei iin* has played a critical role in the care of elderly Japanese people. Hasimoto says,

> Needless to say, one of the most extraordinary systems in the field of welfare in Japan is the *Minsei iin* community volunteer network. With a history going back to the beginning of this century, or even further, the network currently boasts nearly 200,000 members all over the country. The *Minsei iin* volunteers are said to be an irreplaceable link between the authorities and residents, people in need. It is the most brilliant example of neighbourly love and responsibility turned into help in this eastern nation. (1995: 190)

While *minsei iin* remains an important presence, the practice might change now that Japan has realized its need to develop community care to cope with population ageing. The earthquake experience has drawn even more attention to this need.

Given the urgent need to respond to the increased demand for health care in an ageing population, Japanese research institutes have sought to learn from examples in European countries such as the UK, Sweden and Germany. Reports about the experiences of European countries are discussed at the national level. All focus on health care systems and financial provision; none consider the views of the people who receive the services. The English word 'community', rendered in *katakana*[12] is now used frequently in Kobe. It is one of the most common words in research on local forms of interaction and care, but as a concept it is hard to pin down in the Japanese context. As this book will demonstrate, the Hanshin earthquake prompted more thought about the practice of community care in Japan.

## The Great Hanshin Earthquake

The Great Hanshin Earthquake shook the Hanshin area of Japan on 17 January 1995. The earthquake, measuring 7.2 on the Richter scale, struck the area around Kobe at 5.46 am. It devastated an area twenty kilometers long and one kilometer wide, affecting 1.6 million inhabitants and causing enormous damage to Kobe and other nearby cities in the Hyogo and Osaka prefectures. In recent history, it has been rare for such a powerful earthquake to hit a densely populated and economically important area. This section demonstrates the short-term physical and mental health impacts of the earthquake.

The scale of physical damage was more comparable to disasters in developing countries than richer ones. A total of 400,000 houses were damaged beyond repair and by the 23 January 317,000 people had been evacuated to 1,150 shelters. The total death toll was 6,400; more than half of the fatalities were aged over 60 (Tanida, 1996) and 60% of those killed were women. All of the most common injuries and earthquake-triggered illnesses were reported disproportionately among people aged over 65. According to the 1990 census, people aged over 60 years made up 17.8% of the population in the affected area (Tanida, 1996). In terms of death and morbidity, the elderly population of Kobe was disproportionately affected by the earthquake.

In the aftermath of the earthquake, there was an urgent need to provide housing for resettlement. Within one month of the earthquake, temporary shelters for 10,000 households were built in Kobe and Osaka. By the end of August, 50,000 households were sheltered at no cost to the homeless people. Some temporary shelters included special homes with care for elderly people. In the following years, PRH (public-reconstruction housing) was built in the form of modern high-rise apartment buildings and offered to the disaster-affected people at a heavily-subsidized rent.

Priority was given to elderly people in allocation of PRH. While this could be attributed to the traditional respect accorded to elderly Japanese people, it might also be a statistical measure of need, or a result of media coverage which emphasized the high proportion of deaths and injuries among the elderly. Two PRH compounds were specially equipped with Silver Housing buildings which included facilities and staff for the care of elderly people. One of these compounds features in the ethnography reported in this book. Some elderly residents made permanent homes in PRH schemes, while others chose to move into the private rental housing or shared a group home with others.

Two weeks after the earthquake, it was reported that 4,500 jobs had been lost as small businesses went into liquidation. Medical facilities in the areas were also greatly affected. Of 222 hospitals, 191 received more or less heavy damage and the buildings of 15 were completely destroyed. Similarly, out of 4,578 clinics, 2,479 were more or less destroyed. In addition, the earthquake damaged the lifelines of the city: water pipes, electricity cables, communication lines, gas pipes, and roads. This type of damage significantly hindered rescue efforts, thus worsening the disaster.

News reports presented the earthquake as not only a natural disaster, but also a man-made one. Although on the surface Kobe appeared before the earthquake to be a very modern, westernized city, much of the less affluent people lived in urban areas where many old houses were built of wood. In contrast, the middle-class tended to live in residential areas on the mountain side of Kobe, as well as in neighboring cities such as Nishinomiya and Ashiya. The downtown area closer to the bay was more crowded. Furthermore, many of the houses destroyed by the earthquake were built prior to the Second World War, having avoided the heavy bombing that destroyed most urban areas at the end of the war. Most of the older houses had seen little renovation, whereas the houses on the mountain side were newer and renovated more frequently, a common practice in modern Japan. Therefore, the old houses were in areas lagging behind in the post-war development.

### *The government response*

In response to the earthquake, the central government granted JPY 900 billion (seven billion US dollars) to prefectures and cities to help rebuild homes and repair damaged services , in addition to the amount spent on rebuilding major city infrastructures such as highways and long distance bullet trains to the Hanshin area. Despite this substantial contribution, the central government was criticized for not offering direct subsidies to individuals affected by the earthquake.

Although Kobe City is the capital city of Hyogo Prefecture, its population of more than one million gives it the status of *Seirei Toshi*, an independently administered city that reports directly to the central government. However, under the disaster relief law Kobe City was subordinate to Hyogo Prefecture, increasing the already existing tension between Hyogo Prefecture and Kobe City and possibly impeding the response to the crisis. The influence of this conflict on the dynamics of the recovery process will be examined throughout this book.

### *Post-earthquake housing schemes*

This book focuses on two kinds of housing schemes that were initiated to respond the earthquake's destruction of around 400,000 homes: TSH (temporary shelter housing) and PRH (public reconstruction housing). Most TSH was *nagaya*-type housing[11], while PHR is generally located in high rise modern apartment buildings. While waiting for TSH to be built, many people found other places to live. The 50,000 people who were too poor to resettle without public support were relocated to PRH over the first five years following the earthquake. The *Asashi Shinbun* newspaper reported in 2002 that the total number of PRH households was 38,600.

## The research aim

The focus of this book is on the resettlement of elderly people with no functioning family after the Great Hanshin Earthquake, first in TSH and then in PRH. This group is of particular interest because their situation drew the attention of the public to the issues associated with the ageing society. It begins by considering the processes of community reconstruction and the resettlement of elderly people after the earthquake. It then examines the potential difficulties that elderly people faced as they were resettle in public housing schemes, particularly in relation to the experience of community and identity. Important in ameliorating such experiences were the skill sets employed by community workers. However, some carers were more skilled than others and there is not a lot understood about what kind of skills were needed in order to respond to the crisis.

The book maps the complexity of the experiences of elderly people after the earthquake and explores the areas where the welfare system did not match the needs of elderly people. It focuses on a specific natural disaster and as such is limited in the extent to which its observations can be generalized to similar events. However, while each crisis might be unique in some ways, other events might share common issues.

## The structure of the book

Chapter two presents a review of the relevant literature, especially comparing Japanese notions of community with similar ideas from other countries. Chapter three outlines the methodology, as well as the limitations associated with it. Chapter four offers a reanalysis of the Hyogo Prefecture post-earthquake health surveys from 1996 to 1998, emphasizing their implications for this study of the experience of elderly people affected by the earthquake. It also quantitatively analyses media data. This data will be followed by qualitative analysis of two and a half years of media coverage. They are therefore not only analyzed for differences in tone and emphasis, but also for variations over time. Gender is a component in the analysis.

Chapters five and six compare TSH and PRH schemes. These chapters utilize discourse analysis to analyze materials collected from media sources, expert opinions and stories from informants. Chapter five examines the process of TSH allocation, arguing that reconstruction brought simmering problems between low-income earners and the elderly to the surface. Chapter six focuses on the public perception of high-rise PRH in both its general and silver forms, emphasizing its impact on the development of community care and health expectations as well as its implication for household structure and relationships. Based on ethnographic research, chapter six presents the dynamism of the reconstruction process in each community, including the attitudes of informants about whether reconstruction was a success or failure. Chapters seven and eight focus on this measure of success by closely analyzing the meaning of two key themes: loneliness and the phenomenon of *kodokushi* (isolated death). These themes are particularly powerful in the way that they linked media, Japanese culture and the problems of service provision.

The concluding chapter sums up the principal conclusions of the study. It discusses critical gerontology and identifies questions for future research. It also summarises the way in which this book has made an original contributions to knowledge about this topic.

# *Chapter Two*
# *Natural Disasters in Literature*

## Introduction

2005 was a bad year for natural disasters: the Indian Ocean tsunami which devastated the coastline of southeast Asia on 26 December 2004, the Fukuoka earthquakes of the 20 March and 20 April 2005, hurricanes Katrina and Rita which hit the southeast coast of the United States from August to September, Hurricane Stan in Central America on 4 October, and the earthquake on the border of India and Pakistan on 8 October, just to name a few. While large-scale natural disasters are not new phenomena, they now receive international attention and governments are increasingly alert to their social dimensions and their implications for development (UNDP, 2004). For example, the massive Wenchuan Earthquake that shook Sichuan on 12 May 2008 (see Table 2.1) received international attention because it occurred during the lead up to the 2008 Beijing Olympic Games. Before the earthquake, the area was not well known outside of China; the earthquake drew international attention to this area's vulnerability to earthquakes as well as to its development. Other magnitude 6 level earthquakes occurred in Xinjiang 2008 and in Yunann, China, in 2009.

There have been heat waves such as those that occurred in 1995 Chicago, 1998 Shanghai, 2003 France and other European countries. Natural disasters bring to the surface the most poor and vulnerable groups in society: orphans elderly and the disabled (Otani, 2006 and 2010, Burrows, 1997, Krause, 1987, Massey, 1997, Phifer et al., 1988, Weintraub et al., 1999). They make visible the previously invisible, such as migrant workers who are not otherwise officially registered (Wilson, 2000, Fletcher et al., 2005). However, while international aid and media attention peak in times of disaster, they are not usually sustained for long after the immediate emergency period and many victims have a great need for support in the long term. Figure 2.1 illustrates the changing health needs after earthquakes and other natural disasters such as tsunamis over a period of time.

While there is extensive medical and epidemiological research undertaken at times of crisis, such studies rarely go on to study the long term effects of a disaster. Little research examines the challenge for victims of rebuilding

Table 2-1: Comparisons of the Wenchuan, Great Hanshin and Fukuoka West-off earthquakes

| | **Wenchuan** | **Great Hanshin-Awaji** | **Fukuoka West-off** |
|---|---|---|---|
| Time | 12 May 2008 | 17 Jan 1995 | 20 March and 20 April 2005 |
| Date | 14:28 | 05:46:52 | 10:53:40 and 06:00 |
| Magnitude | 8.0 | 7.3 | 5.9–6.57 |
| Epicentre | Wenchuan, Sichuan Province (70 km from Chengdu) | North of Awaji Island | Fukuoka West-off |
| *Causalties* | | | |
| Dead | 69,299 | 6,427 | 1 |
| Severely injured | 374,219 | 43,792 | 1,069 |
| Missing | 18,195 | 3 | n/a |
| *Buildings damaged* | | | |
| Total | 216,000 buildings | 460,000 households | 375 households |
| Severe | 4150,000 buildings | | 9,000 |

long after international support has ended. More studies should analyze the psycho-social conditions of victims, community wellbeing and the redevelopment of infrastructure and livelihoods over the long term.

This chapter comparatively reviews previous literature on natural disasters, focusing on their implications for research on the recovery process after the Great Hanshin Earthquake that shook Kobe in 1995. It will draw on literature by Klinenberg (2002) on the 1995 Chicago heat wave, by Morrow (1999) on Hurricane Andrew in 1992, and Taylor's community studies of natural disasters, including the 1995 Great Hanshin Earthquake. The review will focus on the: health impact of disasters, the role of the media in social policy, long-term community development and other social issues.

## Health impacts

This section reviews journal and newspaper articles about the effects of the Great Hanshin Earthquake on the health of victims, with some comparison with relevant studies from natural disasters in other countries.

Figure 2-1: Timing of health needs after an earthquake

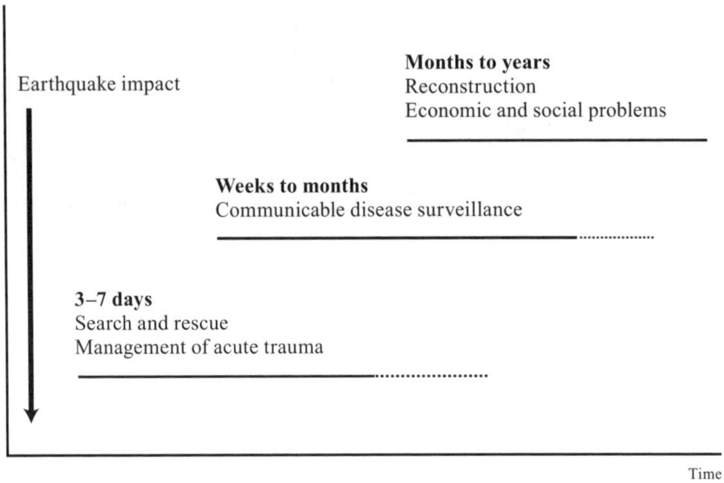


With the exception of the Hyogo Health Surveys that will be discussed in chapter four, most articles drew from case reports. According to Noji (1997a and b), post-disaster research aims to assess the needs of disaster-affected populations, efficiently matching resources to needs, preventing further adverse health effects, evaluating program effectiveness and carrying out contingency planning. However, this rational-scientific approach is difficult to measure in emergency situations and has rarely included elderly people as a specific group for analysis (see for example Noji et al., 1993 1997a and b, Goenjian et al., 1997, McDonnel et al., 1995a and b and Armenian et al., 1992, 1995 and 1997 for research on disasters on the American continent and in Armenia). Researchers have utilized a non-epidemiological approach to analyze disaster areas in Italy, Turkey, Iran, India, the Philippines, Australia, Korea, China, Mexico, Guatemala, Nicaragua, and Egypt as well as the USA and Armenia (Carr, 1997a, b, and c, Karanci and Rustemli, 1995, Lima et al., 1992a and b; Noji 1997a and b; Vanholder et al., 2001). Many of these studies were done in collaboration with universities in the United States or the United States CDC (Center for Disease Control and Prevention) (Noji et al., 1993). Most of these papers focused on 1995 Great Hanshin Earthquake in Japan, the 1994 Northridge Earthquake in California, the 1988 earthquake in Armenia and Hurricane Andrew, which hit the United States in 1992.

While most of the studies of the Great Hanshin Earthquake have focused on elderly people, studies of the Armenian Earthquake and Hurricane Andrew focused on children and adolescents. Research on the Armenian

Earthquake generally focused on post-traumatic stress disorder (Armenian et al., 2000), especially among children (Goenjian, 1993, 1997, Pynoos et al., 1993, 1998, Najarian et al., 1996), and was conducted in the first eighteen months after the earthquake (Goenjian, 1995). Goenjian (1994) looked at some elderly cases, but the bias towards the study of children likely resulted from the fact that two-thirds of the victims of the Armenia earthquake were under 12 years old (Azarian et al., 1996, Miller et al., 1993).

The Great Hanshin Earthquake caused more than 10,000 deaths, many more than earthquakes in developed countries of a similar magnitude. Usually, major earthquakes that occur on the west coast of the United States and in Australia are not listed in public health articles because they result in less than ten or at most a few dozen deaths. The reasons for this difference are related to population density and housing conditions rather than the degree of industrial development *per se*.

Few studies have taken into account the different health consequences of a disaster for people based on their socio-economic background and other social factors. However, this approach is helpful because different social groups tend to be affected in different ways. For example, many Japanese articles and books about the short-term health effects of the 1995 earthquake drew attention to stressful living conditions in evacuation centers (Kobe University, 1997a and b, Iwasaki et al., 1999). Many survivors were located in large facilities with no privacy, such as a crowded school gym. It was very cold as the earthquake occurred in January, and emergency toilets were often built in the open air places like school yards. Subsequently, many of the elderly residents restricted their eating and drinking because they did not want to wake up in the middle of night to walk to the uncomfortable toilet outside in the cold, potentially disturbing other people nearby. This caused many to suffer from dehydration and malnutrition (Tanida, 1996).

Another limitation of previous studies of disasters is that they have generally only investigated the period immediately following the disaster. Studies of short-term psychological damage are more numerous and extensive than long-term ones, even though researches often suggest the importance of conducting follow-up studies if they are concerned about latent health problems, the delayed onset of symptoms, or the likelihood that symptoms wax and wane (Bland et al., 1996). For example, Gleser et al. (1981) found that survivors who were relocated after the Buffalo Creek disaster in New York showed psychiatric symptoms for as long as 14 years (Bland et al., 1996). Outside the area of psychology, some suggest disaster survivors are at greater risk of mortality and morbidity from diseases such as coronary heart disease (Melkonian et al., 1997).

Studies of the long-term effects of disasters are more difficult to carry out because public attention fades over time and it can be difficult to identify and reach the affected population. Many move on with their lives, leaving temporary shelters and resettlement communities and changing their addresses in the course of reconstructing their lives.

Long-term problems can be economic as well as health related. Kobe News reported that those who lost their businesses at the time of the Great Hanshin Earthquake had managed to reopen it after one or two years. However, by the third year many went into decline and some had closed down due to community wide suffering and the decreased population.

There is no epidemiological survey of the psychical wounds caused by the earthquake, but reports have outlined six common clinical diagnosis: (1) injuries such as crush syndrome[1], spinal fractures[2], renal replacement[3], acute renal failure[4], and burns; (2) injuries to the circulating system such as hypertension, high blood pressure, and coronary heart disease; (3) injuries to the digestive system such as hemorrhagic gastric ulcers and peptic ulcers[5]; (4) injuries to the respiratory system such as pneumonia, and bronchial asthma; (5) injuries to the metabolic system such as diabetes mellitus; and (6) other injuries including mental stress-triggered recurrent endogenous uveitis. All have shown a disproportionately large number of cases reported in elderly people. The Kobe University Hospital reported the worsening conditions of all the above kinds of disease suggesting their probable association with the earthquake (Yamamoto and Mizuno, 1996).

A report from the Kobe University hospital showed that among other injuries, crush syndrome[6] was particularly common among elderly victims of the earthquake, in addition to the expected injuries to arms, legs, spine and head (Yamamoto and Mizuno 1996, Oda et al., 1997). A group from Osaka University Medical School reviewed the medical records of patients admitted to 95 hospitals within or surrounding the affected area during the first 14 days after the earthquake (Kuwagata, Oda et al. 1997). Figure 2.2 shows the age and sex distribution of those 230 patients who agreed to be interviewed, 140 with spinal fractures and 100 with rib or pelvis fractures. The average age was 62.9 years old. As for the sex distribution, 70% (162) of them were elderly females.

This study is compared with the report from the 1976 earthquake in China, which resulted in 242,769 deaths and 164,851 injuries. The three major injuries there were crush syndrome and fracture of the pelvis and of the spine. Among the cases of spinal fracture, 70% sustained injuries of the thoraco-lumbar spine and 34% were paraplegic. This high incidence, it is thought, may be due to the collapse of Chinese stone houses, while in Japan

Figure 2-2: Age and sex distribution of major bone injuries

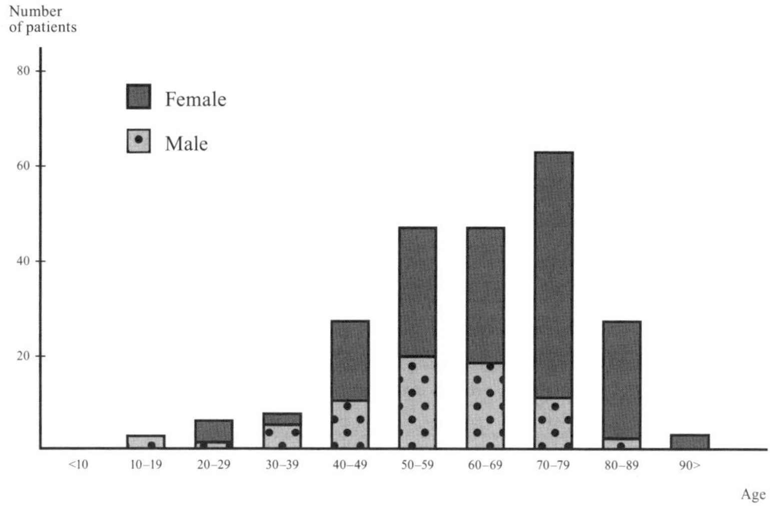


Source: Maruo and Matumoto (1996)

houses are wooden. In comparison, the 1988 Armenian earthquake resulted in 25,000 deaths and 30,000 injuries and reported few spinal or other trunk fractures (Burk, 1993). In Japan, 995 (59%) of 1675 reported bone injuries were fractures of the spine or other trunk bones, although only 21 (2.1%) had a spinal cord injury. It is suggested that this is due to the Japanese traditional custom of sleeping on futon mattresses on the floor because most were injured when they were getting up from the floor (Shimazu et al., 1997).

In the Great Hanshin Earthquake, 504 deaths were listed as fire related, although many of the victims may have been crushed or suffocated before they were burned (Nakamori et al., 1997). Among the 2,718 patients of the above medical record review, 44 patients (1.9%) were hospitalized with burns. Figure 2.3 shows the age and sex of the 44 patients. Morbidity increased with age and was higher in patients over 40 years old when calculations were based on the population in the affected area.

The Kobe University Hospital observed an increased morbidity from diseases of the circulatory system such as hypertension, high blood pressure and coronary heart disease, suggesting that this was associated with the excessive stress and the hard work of the emergency (Yamamoto and Mizuno 1996, Kario et al., 1997a and b). Yoshikawa (1995) reported cases of cardiac emergencies after the Great Hanshin Earthquake. Congestive heart failure occurred mainly in old patients who suffered from upper respiratory infection

Figure 2-3: Age and sex distribution of burns

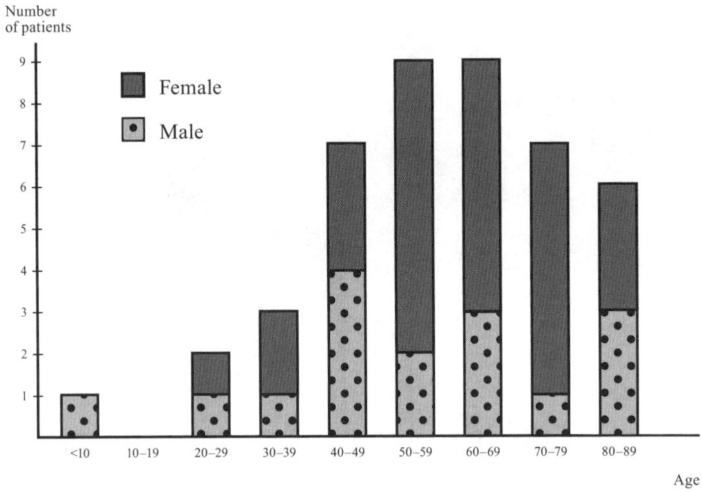


Source: Nakamori, Tanaka et al. (1997)

or bronchitis, triggering heart failure. Another problem was that they were not able to take their regular medicines under emergency conditions. Their average age was 70. The high morbidity from angina could have resulted from the cold and stress of the evacuation (Yoshikawa, 1995).

A link between emotional stress and sudden cardiac death was also reported in the data from the 1994 earthquake in Northridge, California. On the day of the earthquake, there was a sharp increase in the number of sudden deaths related to atherosclerotic cardiovascular disease. During the six days after the earthquake, the number of sudden deaths declined to an average below the base-line value. Leor et al. (1996) concluded that the earthquake was a significant trigger of sudden death due to cardiac causes, independent of physical exertion.

The morbidity of diseases of the digestive system, such as hemorrhagic gastric ulcers and peptic ulcers, rose with the increase in mental stress and the worsening of living arrangements and eating habits, according to Kobe University Hospital and other reports (Yamamoto and Mizuno 1996, Takakura et al., 1997). The hospital also reported an increase in diseases of the respiratory system such as pneumonia and bronchial asthma a few weeks after the earthquake. This could be explained by the dramatic changes of environment resulting from the occurrence of the earthquake at the coldest time of winter (Yamamoto and Mizuno, 1996). Takakura et al. (1997)

outlined the diverse influences of the Hanshin earthquake on pneumonia and bronchial asthma. Following the initial rush of victims with surgical and orthopedic problems, patients with respiratory diseases increased within one month of the disaster, particularly among the elderly.

Mental health problems could be divided into three categories: short-term mental disorders caused by earthquake shock, the triggering of latent diseases triggered by stress, and symptoms caused purely by the earthquake, including PTSD (post-traumatic stress disorder). In the second category, the most frequently reported were dementia triggered by mental stress and environmental change and alcohol-dependent syndrome. Eczema and allergy were reported both as a result of mental stress and the related decreased immune system.

Although Japan is often subject to natural disasters, the psychological effects on disaster victims have not been widely studied. Only a few have investigated the effects of a volcanic eruption (Kato et al., 1996). However, after the Great Hanshin Earthquake, many studies have been conducted on post-traumatic symptoms and PTSD, a western term that became widely used in Japan following the earthquake. Many books on psychological counseling have been published since the Hanshin earthquake. In Japanese society, people have been reluctant to seek psychological help, unlike in the USA. It is seen as acceptable to seek other medical treatment but there is still a relatively high barrier against going to a psychiatrist.

Most studies of the psychological consequences of disaster tend to focus on the long-term effects rather than the short-term, while most reports on other physical health effects tend to be on the short-term rather than the long-term. Studies of the psychological effects have been reported from earthquakes in Ecuador (Lima et al., 1992a and b, 1993), Armenia (Goenjian et al., 1994a, Goenjian et al., 1994b, Louis et al., 1996), San Francisco (Cardena and Spiegel, 1993), Italy, (Bland, et al. 1997), Turkey (Karanci and Rustemli 1995), Australia (Carr et al., 1997a, Carr et al., 1997b and Carr et al., 1997c), India (Sharan et al., 1996) as well as many from other natural disasters such as hurricanes in the United States (McDonnell et al., 1995b).

Research on the impact of disaster on mental health in old age has given conflicting results. The elderly may be best prepared for disaster because of their previous life experiences, or they may be more vulnerable than younger age groups because of their frail health, lower adaptability, strong emotional attachments to long cherished property and lost mementoes, or because they tend not to proclaim their problems unless they are asked (Tanida 1996, Gerrity and Flynn, 1997). Different studies have reached opposite conclusions. Robertson and Bell (1976) concluded that the old and poor were reluctant to use available resources and the shock of a natural disaster seemed to last longer among elderly individuals. However, other findings of research on post-traumatic symptoms after the Hanshin earthquake concluded that the elderly were better protected from the stresses of disaster when compared with the young, contrary to the expectations of the research. At the first assessment in the third week, subjects younger than 60 years old and subjects older than 60 years old experienced sleep disturbances, depression, hypersensitivity and irritability. During the second assessment in the eighth week, the percentage of younger subjects experiencing symptoms did not decrease, while elderly subjects showed a significant decrease in eight out of ten symptoms (Kato et al., 1996).

Kato et al. (1996) gave three explanations for their findings. Firstly, the younger evacuees may have experienced greater psychological stress in reconstructing their lives and the lives of their families and in finding new jobs than the elderly who were retired and receiving a pension. From my personal conversations with the elderly people in the shelters, they had been living on their pensions and personal savings, or as recipients of livelihood protection support, so the need to find a job was not a problem for them either before or after the earthquake. Secondly, the elderly might have established better social networks in the shelters than the younger survivors because they had lived in the local community for longer before the earthquake. The third explanation relates to previous disaster experiences. Those older than 60 years at the time of the earthquake were born before 1935, so had lived through the adversities of the Second World War, including the severe destruction of cities by heavy bombing, and then through the post-war rapid reconstruction and economic development. The destruction of the area by the earthquake was often compared with the situation at the end of the Second World War. The younger subjects were experiencing a large-scale disaster for the first time, so recovery from the psychological impact may have been delayed compared to more experienced, older subjects. The older people in my research also mentioned that this was their third experience of large-scale disaster, as there was also the Kobe flood in 1917. However, Kato et al's study was limited in its interpretation of the data and absence of controls. Symptoms such as sleep disturbances and irritability do not necessarily predict the degree of psychological impairment.

Finally, Knight et al. (2000) have pointed out that mental health problems following a disaster cannot be solely attributed to the disaster. Their study of the mental health of older people after the 1994 Northridge Earthquake showed that prior earthquake experience was related to lower post-earthquake depression scores while the strongest influence on post-disaster mental health was pre-disaster mental health. In Kobe, it is possible that those survivors who predominantly lived alone and on a low income before the earthquake were already suffering from depression related to loneliness and isolation before the earthquake.

Although there are more articles discussing the consequences of disaster for mental health rather than other health consequences, there are critical gaps in knowledge. In particular, loneliness and isolation may be long-term mental health problems (Gerrity and Flynn, 1997). In a study of the 1997 earthquake in Umbria, Italy, people living in prefabricated huts showed a higher score on the Geriatric Depression Scale and on the Hamilton scale for anxiety and complained more often about their health status compared with those living in permanent buildings. While all of the participants suffered from the discomforts caused by the earthquake, the precariousness of living in temporary houses could help to explain the higher distress experienced by those housed in the huts (Mecocci et al., 2000). This would be consistent with what was observed among people living in TSH in Kobe.

## Media and social policy

It was clear from an early stage that the influence of the media would be important in any study of the aftermath of the Hanshin earthquake. The Japanese media provide ongoing commentary on population ageing and their output on this topic was greatly increased after the earthquake. No Japanese major newspaper has a single day without an article on an ageing-related topic. Usually there is more than one. The provincial press and TV were able to maintain interest in survivors many years after the disaster partly because of the age of victims and survivors. Public attention to issues relating to population ageing is very high. Articles about health also appear in the media daily in Japan (Lock, 1996: 208). Despite this, little has been written on the role of the media in social policy. Campbell has shown that the media was one of the main actors in health policy change in ageing Japan (Campbell, 1992 and 1996) but this was pioneering work and more needs to be done.

General social policy books have not discussed the media,[7] nor did the fourth edition of *Older people in modern society* (Tinker, 1997). The textbook, *Social Policy*, edited by John Baldock et al. (1999: 498–508) did not have a media chapter, but had a section on the media.[8] The points are set out below:

> The mass media are crucial to social policy for a variety of reasons, not least because they often perform an influential role in framing many social policy

> debates. With the advent of television we are exposed to the mass media more than ever before. In this way, the mass media have become crucial to how we understand the world, and it is perhaps not surprising that some mass media content has attracted concern and condemnation for being offensive and inflammatory (498).
>
> There is often something of an implicit assumption on the part of involved agencies and commentators that media coverage is important for changing public perceptions and helping to change and modify policy. But is this really the case? What kind of impact does the mass media have upon public attitudes towards different social problems? The first point, of course, is to recognize that the media are far from homogeneous (499).
>
> It is important to to recognize that the media operate with their own agendas (499).
>
> The ability to reach and sensitize so many people to social policy concerns is not to be minimized. Yet the assumption that media coverage of a social issue will have a direct impact upon both public opinion and policy-makers is a questionable one (499).

Following on from this last point, Klinenberg (2002) noted that journalists themselves disagree about the kinds of roles they should play in reporting public events and what policies to follow when deciding what is news.

Although the role of the media was not addressed in general social policy books, there are some books that are focused on the media and social policy (Pharr and Krausseds, 1996, Franklin, 1999, Rubinstein and Brown, 1985). Franklin described twenty cases of media influence. One case showed that the unrelenting press criticism and misrepresentation of social work has had important consequences. It has demoralized social workers, influenced their professional practice and, by helping to shape public opinion, impacted ultimately on social policy concerning social workers and their clients (Franklin, 1999: 3). While the media appear to have had a negative effect on the profession of social work in the UK, it is possible that they had a more positive effect of the development of the profession of Life Support Advisor in Japan (see chapter six for this discussion).

In Japan, the media have an important influence on public opinion formation (Campbell, 1992). This was supported by this research; when interviewed officials mentioned the media as influential in forming people's views. It may be particularly important to elderly people. Anne Cooper-Chen found that Japanese elderly women watch television for 6.5 hours a day, while men in their sixties and in their seventies watch four and five hours, respectively (Cooper-Chen, 1997: 106). She concluded that television in Japan provided social welfare services in the absence of other activities for the elderly (Cooper-Chen, 1997: 127). Interviews with elderly people also showed that the can media shape the way that elderly people see themselves and how they are seen by others (Biggs, 1993). Klinenberg stresses that these images and stories have to be selective. The nature of news requires fast thinking; on breaking news stories and journalists face intense time pressure (Klinenberg, 2002: 210).

Another important feature of news lies in the selection of headlines and visual images. Headlines are important because they present news in a way that allows for selective reading, and because they suggest which events and issues matter most. Few people have time to read an entire newspaper, but many scan its headlines and photographs as part of their daily routine, reading the details of only those items that catch their interest (Klinenberg, 2002: 213).

Klinenberg also makes the point that representation in media coverage does not necessarily reflect the real world. Editors and producers are always looking for dramatic images for the lead story (Klinenberg, 2002: 217). For example, in post earthquake coverage the word *kodokushi* (isolated death) was a popular headline (see chapter eight for a discussion on why *koodokushi* particularly interested Japanese readers).

In 2002, too late for the research design of this book, Klinenberg published *Heat wave: a social autopsy of disaster in Chicago*. This work stressed the importance of the media in responding to a disaster. Disasters are a journalistic staple that rank among the most popular forms of news content. Klinenberg suggests that the media were important for two reasons. One was that the media had created a fear of crime among older people which stopped them from going out. This fear of crime, which elderly people heard about daily on the radio and television, therefore contributed to their confinement (Klinenberg, 2002: 51). Since Japanese society has an exceptionally low crime rate by international standards this was not an issue for the elderly earthquake survivors. However, fear of unscrupulous sales people was reported as preventing elderly people from opening the door to strangers in the new PRH schemes, even if the visitor was a potential helper.

The second concern over the media arose from conflict with the government's public relations. According to Klinenberg, the Chicago city administration accomplished an excellent public relations campaign to deny the severity of the crisis, deflect responsibility for the public health breakdown and defend the city's response to the disaster (Klinenberg, 2002: 168). The campaign to manage the disaster had an impact on the journalistic coverage of the heat wave (Klinenberg, 2002: 184). Again this did not appear to be a problem in Japan. While there were tensions between local government and local communities that were not highlighted in the media coverage, no systematic defensive response by government was discovered in my research. The message 'government alone cannot do it all' was not just an excuse from government, it was a reality.

Earlier sections of this chapter showed how the influence of the media is largely ignored in social policy, highlighting the importance of media in Japanese social policy. The contribution of the post earthquake media coverage highlighted the increasingly important group of impoverished elderly people living alone in urban environments. A careful examination of media images that are usually taken for granted can help us to reflect critically on received knowledge (Cirillo, 1994: 173). Discourse analysis of media data media can show how the real world and the media interact, revealing changes in the media reporting and analysis and the varying degree of influence on social policy over time (Liddiard, 1999). Despite the obvious importance of the media in Japan, Takayose (1999) wrote that when social problems occur, academics and the media concentrate on emotional critiques rather than aiming to improve policy implementation.

## Disaster as a social issue

Most literature on natural disasters has taken an epidemiological approach, even if the full rigor of epidemiological studies has rarely been achieved. Such studies also rarely focus on elderly people. One exception is Klinenberg's (2002) study of deaths in the 1995 Chicago heat wave. As he says, US epidemiological reports on disaster generally identify a relationship between morbidity, mortality and socio-demographic factors, but they offer little explanation as to why and how issues are related (Klinenberg, 2002). Epidemiological studies showed that in the 1995 Chicago heat wave social contact was a key factor in determining heat wave vulnerability. Klinenberg examined which social conditions facilitated strong and effective support networks, and which conditions rendered frail residents even more susceptible to deprivation and isolation (Klinenberg, 2002: 33). Such social contact, called *tsunagari* in Japanese, is what volunteers in Kobe like the *minsei iin* were working to achieve for earthquake survivors, especially elderly people. The word *tsunagari* often featured in headlines and news broadcasts about the earthquake recovery.

Morrow (1999) has proposed an approach for analyzing the social aspects of disasters. She sees the vulnerability of victims to disasters as socially constructed; that is, it arises out of the social and economic circumstances of everyday living. In her study of Hurricane Andrew, she identified certain categories of people as being at greatest risk during a disaster response: the poor, the elderly, women-headed households and new migrants. These are socially vulnerable groups even in normal times. Taking an area based approach she found that knowledge of where these groups were concentrated within communities and the general nature of their circumstances was an important step towards effective emergency management. Morrow argued that the household was an important unit in combating vulnerability as it possessed different human or personal resources such as health and physical ability, relevant experience, education, time and skills (Morrow, 1999). As a group, it is safe to assume that elderly residents are more likely to lack the physical and economic resources necessary for effective response to a disaster and are more likely to suffer long-term health-related consequences, even though elderly households vary with age, health, family and economic circumstances.

Disasters can also make visible what is invisible in the everyday world (Varley, 1994), but society, the media in particular, will choose what aspects to take up. In Kobe, the media concentrated on family, community and the potential for loneliness, but chose largely to ignore poverty. In Chicago, even though the media content was first managed by the city authorities, reporting and subsequent actions drew attention to impoverished elderly people who died alone despite having lived in an affluent urban environment like North America.

### *Family change and vulnerability*

Weakening family ties is one cause of elderly vulnerability that may not be fully recognized until there is a disaster. In Japan, research has tended to assume strong family ties (Shanas, 1979a and b). Hashimoto (1996) investigated attitudes of happiness and independence among older women, emphasizing family relations but not community links. Yamamoto and Wallhagen (1998) focused on family relations between mothers and care-giving daughters-in-law as one aspect of family life. Research on the family with a gendered perspective will be an expanding area of research in the future as traditional attitudes weaken. For example, Morrow (1999) found that although relatives were not likely to be the primary source of assistance in developed nations, they were an important base of disaster-related help for many (Kendig et al., 1991, Haines et al., 1996). In Japan, although relatives can be said to still take more active roles in daily assistance than in other developed countries, the earthquake showed that many survivors were lacking family support. This was a more serious problem than it might have been in societies where assumptions about family support are weaker and formal sources of support are better developed.

Few academic studies of community care, or of the general importance of community for the elderly, have been undertaken in Japan (Ninomiya, 1989), but media attention has been increasing in recent years. Yazawa and Kunihiro (1999) have pioneered survey research in the area. Ethnographers and qualitative researchers have concentrated on the personal and familial level rather than the community. Japanese work on retirement communities and group homes is also limited (see Hatoyama and Yamai, 1999) In addition, English language work is only slowly being translated. For example Ungerson's *Policy is Personal* (1987) was only translated into Japanese in 1999.

Morrow (1999) found that people responded differently to a disaster depending on their gender. In the case of Hurricane Andrew, these differences were especially apparent in household preparation, evacuation, and in the use of social and family networks. The research area was Texas, in the south of the United States. Texas is more conservative than northern states, and this would likely be accentuated in times of crisis. Morrow noted that it was important to consider women's care-giving responsibilities, both within households and in responding organizations (Morrow, 1999). Sexual stereotypes and expectations which profoundly influence the daily lives of women and men are even stronger in Japanese society. Informal care is largely provided by housewives. Volunteer group members have been largely middle-aged full-time housewives, or those who work part-time only. These roles resulted in many female survivors having larger networks to fall back on. This gender difference has been observed in many societies (Alpass and Neville, 2003, Perren et al., 2003). Japanese society is not an exception (Otani, 2000a and b, Yazawa and Kunihiro, 1999).

In Kobe, as in Chicago (Klinenberg, 2002), the media drew attention to the plight of elderly men who appeared more vulnerable than elderly women. Klinenberg argued that men were at greater risk during the Chicago heat wave. The elderly, especially isolated men and those who have outlived their social networks or have become homebound and ill, often suffer from social deprivation and role displacement in their later years. Older women are more likely than men to be poor, sick, and living alone in old age, but they also tend to be less isolated. This does not mean that they do not have problems associated with ageing alone (Klinenberg, 2002: 230). Further, as noted above, the media drew public attention to elderly people living alone, presenting them as vulnerable. There is a strong cultural bias against elderly people living alone in Japan so this was not surprising. Subsequently, these older people were considered to be one of the groups most needing attention and support, which they received from the local department of health and welfare.

### *State and community involvement*

Although the destruction of housing and infrastructure was the most visible result of the disaster, there is more to housing loss than the architectural fabric. The destruction of homes also affected community ties and networks. For most elderly people in more affluent parts of the city, their house was probably their major financial asset and main barrier against poverty (Moser and McIlwaine, 1997). But in the Nagata Ward of Kobe City, the proportion of the population who received public assistance before the earthquake was above average. This ward, with a very high concentration of disadvantaged people, was the ward most affected by the earthquake. Many survivors from this area, concentrated in TSH, were among the most deprived; ten percent started to receive public assistance after they moved into TSH. A high proportion[9] of those who moved from TSH to PRH had been recipients of welfare since before the earthquake (Doshisha University, 1997: 12). This can be contrasted with the fact that in 1994, only 0.7% of the Japanese population was in receipt of some form of public assistance (MOHW, 1996).

Within the existing welfare system, some heads of expenditure rose due to increased demand after the crisis. TSH and PRH played a role in mitigating the poverty of disaster-affected people, but the numbers of recipients of unemployment insurance and of public assistance increased in the year following the earthquake. Central government pays for public assistance although it is managed by the welfare offices of local government.

Free medical care was available to people affected by the disaster in the immediate aftermath to the earthquake. Emergency care was given until December 1995, but there were no financial subsidies for long-term medical relief. Elderly people living in the affected area were likely to have to make high co-payments for long-term medical relief as well as pay for the increase in transportation costs due to relocation. After their sudden relocation, mostly to remote and inconvenient areas, many older people had to find a new doctor, possibly one that knew little of their health history and with whom they had to establish new rapport. Fieldwork established that many people who could find their old doctors continued to see them even when they had to spend considerable amounts of time and money to visit them. The family doctors (mostly private small clinic practitioners) in the area were themselves affected by the earthquake. Clinics were destroyed and medical records were lost. This contributed to the destruction of local community networks following the earthquake.[10]

Older men (and some women) who had not reached pension age received little media attention over the long term. Many had lost their jobs and their chances of finding new employment to tide them over until they reached pension age were poor. There was no assistance specifically provided for this group. However, since they have now begun to receive more media attention, measures may yet be developed.

### *Housing*

The Meiji Restoration and the process of democratization during the post-war occupation aimed to make Japanese society more equal; class distinctions have not been a major issue as in the UK. Yet there are still cleavages that relate to income and the gap is widening. By the time I started my research, nearly all those in the fieldwork sites were from the most disadvantaged groups of society. They were poor and poverty is isolating (Klinenberg, 2002: 71). Although polarization of housing by income level had existed in Kobe before the earthquake, it became even more visible in the post-earthquake reconstruction process (Hirayama, 2000), implicating government at different levels. Responding to the special needs of funding for housing reconstruction, local governments made loans available to disaster-affected people to help to rebuild their houses (Takayose, 1999: 48). The loans were available to those who already had some money to rebuild their homes. However, for those who had no money, this program provided no help. As public loans did not cover the full cost it was necessary to combine different loans. Some people who were too old to qualify for private loans were able to take out loans in the name of their children. This could only be arranged by those whose children were not themselves affected by the earthquake. This showed that children often function as a safety net in Japan. While rebuilding was expensive for homeowners, tenants found the rent for newly constructed houses was more expensive than before the earthquake. Many people had to give up the idea of moving back to their old areas as prices rose.

Housing-reconstruction policy helps to structure living arrangements in that TSH and most PSH were designed for nuclear families. In the early stages, some extended families in TSH explained that they had to separate, as the space was too small for the whole family. It is therefore reasonable to suggest that the post-earthquake housing situation encouraged an increase in the number of nuclear families among the population and contributed further to the lack of support for elderly people. Elderly men and women who had no functioning family and nowhere to go were housed in TSH. This concentration of deprived elderly people created unbalanced settlements which grew more unbalanced as the years passed. Writing one year after the earthquake, Tanida wrote,

> The elderly people had to live isolated in ordinary temporary houses. Furthermore, because elderly and disabled people were initially given priority in the distribution of temporary houses, this well intended scheme produced a community of elderly and disabled people living alone (1996: 1133–5).

Over the next four years the TSH were slowly emptied and those who could not find other housing were relocated to new PRH schemes elsewhere. The physical condition of public housing in the post-earthquake era was a great improvement technologically on the pre-earthquake housing. This was partly due to the technological advances in the newer buildings and partly because quality regulations for new buildings had become more stringent. It is also partly because attention was paid to putting elderly people in appropriate housing. The concept of housing for elderly people is still relatively new in Japan.

The housing allocation process moved survivors out of TSH and into high-rise buildings; most were wholly or partly filled by elderly people who lived alone. Taylor (2003) has analyzed the UK experience of re-housing communities in new towns and new housing estates outside the inner cities after the Second World War. This transfer of significant populations to new neighborhoods with few facilities and away from their traditional networks led to isolation and created a new set of social problems (Taylor, 2003: 19). In Kobe City, similar problems arose and the local government realized the need to encourage community self-help and local support networks. The intention was that key community workers would encourage residents to help themselves by developing community ties and local activities. They would also build up the capacities of local residents to acquire individual and organizational skills to help them address their problems. The Natsu Aki PRH research site was an example of the re-housing of old communities in new towns and new housing estates outside the inner cities. Such relocation to new neighborhoods led to similar problems of apathy and isolation among the new tenants as Taylor had found in the United Kingdom. However, in the United Kingdom the whole age range was being re-housed, whereas in Kobe it was almost entirely people aged 50 to 90. The limited age range and the concentration of older people made the process of community regeneration much more difficult.

There appeared to be some confusion on the part of both the authorities and the media over whether the aim of staff and leaders in the new housing schemes was to provide support to the residents of TSH and PRH or to enable them to build their own communities. Loneliness may be one of the most important housing issues for older people (Heywood et al., 2002: 58–59), but providing protective and supportive services to people with limited mobility and extreme needs is a difficult job, even for organizations that are explicitly designed for the task (Taylor, 2003: 105). In a Western country with Christian organizations, a neighborhood church may take a role in reaching out to those who are most isolated and vulnerable; they use networks they already know and trust. But local organizations such as the neighborhood church may not be able to do this work effectively unless they have financial and material support (Taylor, 2003: 105). In Japanese society, it is the *minsei iin* who have through tradition gained the trust to take on this role (Takahashi and Hashimoto, 1997). Although the *minsei iin* are unpaid volunteers, they are appointed by a formal selection process. It is an honor to be appointed. The person has to be helpful, thoughtful, sensible, wise and respected. As will be shown in chapter six, the role of *minsei iin* has influenced community regeneration and community work. The work of *minsei iin* involves contacting people and local government and knocking on doors. The presence or absence of *minsei in* activities influenced the selection of my fieldwork sites (see chapters three and six).

Although the role of *minsei iin* was well developed at the time of the earthquake, they were faced with new challenges in the resettlement schemes. The task of rebuilding communities meant they needed to develop new partnership skills, both within communities and between individuals (Taylor, 2003: 198). Taylor notes that it is often assumed that the two will work together in partnerships, but points out that much of the research on community participation suggests a wide ranging lack of the basic capacities for partnership. Her work has led her to conclude that little thought has gone into evaluating the skills that make partnership work. Morrow (1999), on the other hand, found that it was important to recognize women as a seriously under-used resource in disaster management and response, particularly in decision-making roles. She documented women's contributions in a multitude of roles, from grassroots organizing to emergency management. Certainly, the post-disaster phase in Kobe had thrown up new or extended roles for women working in the community. LSA (Life Support Advisors) were mostly women and some of the most successful *minsei iin* were women, although the post was traditionally a male preserve. The increasingly important role of women is an aspect of the new and sustained strength of the voluntary sector in the post-disaster era.

Sustainable development programs require leadership and involvement at the local level. This important notion is inadequately applied in the disaster context (Morrow, 1999). The challenge comes with the need to sustain reconstruction programs beyond the disaster stage. This was recognized as a problem in community generation in PRH. The key is to have leadership from within the community, not by a volunteer from outside the community (Otani, 2000a). This bottom-up approach to community development should lead, according to Taylor (2003) to participation and the empowerment of community members and to the emergence of local leaders who are able to draw on adequate resources. Her theory assumes that poor and fragmented communities have to be developed in opposition to the authorities. The Japanese approach is one of consensus where it is assumed that public administration and public servants are doing the best for people. The main opposition to community development found in new PRH communities came from apathy and dependence, not the attitudes of the community workers. In Japan, the aim was to get people to participate in traditional activities such as tea drinking and clubs. Top down political membership of committees continued because the main structures of power were not expected to shift. The Japanese emphasis on community generation through social duty is in opposition to Taylor's theory of empowerment in conflict situations.

This does not mean that sustained community leadership could not be a problem in the future. As Taylor (2003) notes, successful community leaders were feted, promoted and adopted by public authority partners in ways that were not transparent. A number of factors contributed to this. The first was that public sector partners and other power holders too often selected the community partners with whom they wanted to work. Inevitably these were likely to be those whom they found it easiest to work with, the most easily approachable and the most acceptable face of community involvement (Taylor, 2003: 133).

Sustainable leadership also requires coalitions across neighborhoods that can provide the space for different groups to come to a common view, thus helping to ensure that power holders do not 'divide and rule' by playing one neighborhood off against another (Taylor, 2003: 189). Chapter six will discuss *mimamori* (the watch-with-care town development committee) meetings that illustrate the growth of coalitions between neighborhoods and communities of interest. These meetings, based on consensus rather than conflict, provided an important space where knowledge exchange could be maximized (Taylor, 2003: 200).

### *Isolation*

Like the Japanese media after the Hanshin earthquake, Klinenberg (2002) examines occurrences of isolation and death resulting from the Chicago heat wave. Why did so many die alone during the heat wave? How can this be understood in the context of the many Chicago residents, especially the elderly, living alone with limited social contacts and weak support networks? Klinenberg explored four features of disaster-related urban governance: (1) the delegation of key health and support services to paramilitary organizations that were not designed to deliver them, (2) the lack of an effective system for organizing and coordinating the service programs of different agencies, (3) the lack of a public commitment to provide basic resources, such as health care and energy, necessary for social protection of the vulnerable, and (4) the expectation that frail and elderly citizens will be active and informed consumers of public goods.

Some of these issues were important in the case of the Hanshin earthquake, but more salient were (1) the absence of family, (2) isolated death, (3) the development of community services and (4) public housing during reconstruction. The Hanshin earthquake raised different issues to the Chicago heat wave partly because it was of much larger scale, and because the Japanese government was inevitably mobilized from the start. There are also important social differences between Japan and the United States in general, as well as the cities of Kobe and Chicago in particular. The crime rate in Japanese cities is much lower than in other big cities around the world. The crime rate per 100,000 inhabitants for Japan in 1999 was 1,705 whereas for the United States it was 8,517 (UNODC, 2000). In big cities like Chicago (Susser, 1993), the crime rate is higher than the average for the United States (FBI, 2009).

The population of Chicago is also starkly differentiated along racial lines. Minority ethnic groups were the hardest hit and there were variations in deaths between the African-American and Hispanic communities. Health inequality among different racial groups in the United States has been widely reported (Wilkinson, 2006) and similar inequalities surfaced during the disaster. Japan, on the other hand, sees itself as a homogenous society; traditional social caste differences, migration and other intra-communal issues are very rarely discussed, even today. Although disasters may lead to a new visibility for marginalized members of society, media reports of the earthquake were focused on the presence of poverty and weakening family ties, rather than on any lack of homogeneity in Japanese society.

Pre-disaster levels of social cohesion were very different between Kobe and Chicago. The rapidity of inner city social change, the rise of individualism and breakdown in family ties has not taken place in Japan to the same degree as in the United States. The area of highest earthquake damage was one of social stability where a traditionally impoverished community lived in old and relatively unchanged wooden houses. Like in Chicago, a disproportionate number of the affected population was elderly people living in isolation from their families, but community ties were stronger in Japan. Poverty, isolation and loneliness became long term problems that, according to the Japanese media, signaled that the co residence of generations would no longer be the norm.

The issues of ageing and dying first received attention in Japan before the Hanshin earthquake. The Department of Health and Welfare aired the topic in a report in the early 1990s (Okamoto, 1994). Numbers of people living alone are rising almost everywhere in the world, making it one of the major demographic trends of modern times. However, there are differences in living alone, being isolated, being reclusive and feeling lonely. Klinenberg defined living alone as residing without other people in the household, whereas being isolated was defined as having limited social ties. Being reclusive was defined as being largely confined to the household and being lonely was defined as the subjective state of feeling alone (Klinenberg, 2002: 43). Although these states are inter-related, most 'people who live alone, seniors included, are neither lonely nor deprived of social contacts' (Klinenberg, 2002: 43). This is significant, because seniors who are embedded in active social networks tend to have better health and greater longevity than those who are relatively isolated or reclusive. Nevertheless, elderly people who live alone are more likely than seniors who live with others to be depressed, isolated, impoverished, fearful of crime and removed from proximate sources of support when compared to elderly people who live with others.

Seniors who live alone are especially vulnerable during episodes of acute crisis because 'there is no one to help recognize emerging problems, provide immediate care, or activate support networks' (Klinenberg, 2002: 43). Although, Klinenberg regretted the lack of policy interest in the growing phenomenon of elderly people living alone in the United States, authorities in Kobe had recognized before the earthquake that elderly people living alone were at risk of dying alone, and that they subsequently needed special attention from social services even in normal times (Okamoto, 1994). After the earthquake, the degree of this attention was intensified. Evidence from public documents such as health surveys shows how Kobe faced an extreme case in having to cope with the large numbers of at-risk elderly people who had survived the earthquake.

City residents in the 1995 Chicago heat wave were more vulnerable if they did not leave home daily, had a medical problem, were confined to bed, lived alone, or lacked air-conditioning, access to transportation, and nearby social contacts (Klinenberg, 2002: 80). Anything that facilitated social contact, even membership of a social club or owning a pet, was associated with a decreased risk of death. Living alone was associated with a doubling in the risk of death and those who did not leave home each day were even more likely to die (Klinenberg, 2002: 46). However, after the Kobe earthquake pets were problematic as they were prohibited from public housing because of the possibility of them causing complaints with neighbors. Despite this, community health workers were recorded as offering encouragement to elderly men who decided to keep a dog, as it gave them a chance to go out and to exchange words with their neighbors. TV programs also presented the healing power of pets for people with traumatic experiences.

### *Gender and isolation*

The groups identified as vulnerable by the public administration have some characteristics in common, such as poverty, old age and disability. In most societies vulnerable groups also include large numbers of women (Moser, 1997), elderly women in particular (Arber and Ginn, 1995). In Japan, more attention is given to the elderly and the disabled than to the poor (Somusho, 1997). Awareness of gender is undeveloped and gender issues are not perceived in the same way as they are in the West (see for example the Hyogo Health surveys where the emphasis was on 'old people'; very few of the published tables contain any breakdown by sex). More women tend to fall into the status of living alone and falling into poverty in old age than men, but women tend to be good at establishing support networks while men tend to be more isolated (Jerrome, 1992, Orloff, 1993, Arber and Cooper, 1999, Otani, 2000a, Perren et al., 2003).

As Klinenberg notes, the mortality records of the 1995 Chicago heat wave maintained by county and state offices illustrate patterns of isolation, such as the paradox that although elderly women were far more likely than elderly men to live alone, they were significantly less likely to be cut off from social ties (Klinenberg, 2002: 74). Men who live alone are at a greater risk of being isolated and lacking social support networks (Orloff, 1993, Fischer and Phillips, 1982: 253, Hoch and Slayton, 1989: 128). Klinenberg noted that men in the United States have more difficulty than women in sustaining intimate relationships with relatives and friends (Klinenberg, 2002: 75). The gendered division of labor has delegated most family responsibilities and friendship-making efforts to women, while men developed core relationships in the workplace (Jerrome, 1992 and 1999). When they are no longer capable of working, men often not only lose their habitual identity as breadwinners, but also tend to fall out of their work-based networks and become dependent on their partner's social connections and sources of support (Connell, 1995: 21–27). This pattern is also found in Japan. Widowers and divorced men often suffer from failing physical and mental health after they become single, while divorced women and widows are more likely to gain support from their social networks and suffer fewer health consequences from their status change (Rubinstein, 1986: 20–21). Men also face particular emotional constraints to intimacy and friendship, in part because conventional models of masculinity encourage forms of toughness and independence that undermine the cultivation of close ties. The literature on men who live alone consistently emphasizes the individuality and detachment that mark their experiences (Klinenberg, 2002: 75, Jerrome, 1999).

A number of organizations existed in Chicago to address the problems faced by elderly people living alone and to assist them to form connections to a world that has left them behind. Such services often aimed to help those who feel isolated and lonely—old friends may have passed on or moved away and their social networks may have become attenuated. Some seek companionship and friendship. The organizations try to offer substitutes to an elderly person for the family and friends they have outlived, never had, or from whom they are estranged (Klinenberg, 2002: 52). In Kobe, new small organizations started with the same mission as in Chicago as part of the upsurge in volunteering that followed the disaster.

Although the literature focuses on supporting people in later life, a gendered approach can show that other age groups may be even more disadvantaged. In terms of surviving disasters, there may be some groups that receive less media attention, are not old enough to be eligible for existing services and who may be even more vulnerable than the elderly. For example, the epidemiological data for male victims of *kodokushi* was skewed toward the 50s age group, while for women it was skewed towards the 70s age group (see chapter four), indicating very great stress on the pre-retirement age group among men.

Klinenberg (2002) listed four trends that contribute to the vulnerability of the growing number of Americans who are old and poor: a demographic shift towards an ageing population, a cultural condition related to crime and the coupling of it to a culture of fear, a spatial transformation involving first the degradation, fortification, or elimination of public spaces and the clustering of supported housing arrangements, especially in areas with concentrated poverty, violence and illness, and the tendency of individuals, particularly older, single men without children and with substance abuse problems, to lose crucial parts of their social networks and valuable sources of social support as they age (Klinenberg, 2002: 48). All of these would apply to Japanese contexts except for the fear of violent crime.

The 1998 Commonwealth Fund's report, *Aging Alone: Profile and Projections*, highlighted the general aging of the American society (in Klinenberg, 2002: 49). It revealed that most seniors who lived alone were women, about two-thirds of whom were widows. Class status was a key determinant of isolation and living alone. Two out of every three seniors who are poor lived by themselves, a situation that was dangerous because impoverished seniors were twice as likely as financially stable ones to report poor health, have health-related limitations in bathing, dressing and other daily tasks, and to experience depression at least once a week. The combination of isolation and depression often spins into a vicious circle that is difficult to break; being alone leads to depression, which in turn reduces one's capacity to make contact with others, which then heightens the depression, and so on (Klinenberg, 2002: 49).

## Tsunami and hurricanes

Academic journal articles containing medical case reports on the emergency response generally appear within one year of a disaster (CDC, 2005 and 2006, Nishikiori et al., 2006, Byleveld et al., 2005), as does health policy dialogue (Rosenbaum, 2006, Mattox, 2006, Falk and Baldwin, 2006, Fletcher et al., 2005). HelpAge International (2005b) published a research report on the consequences of the Indian Ocean tsunami for elderly people, but the framework could be applied to other groups. It discusses discrimination, individual rights, livelihood security, welfare and social participation.

In response to the 2005 tsunami, the East-West Center in Hawaii published a long-term, cross-country sociological study of the focusing on human rights and policy recommendations (Fletcher et al., 2005). Other than the bulk of media reports (Bhalla, 2006, Stone and Kerr, 2005), mid to long term sociological research is not yet available for sharing. Academic journals are calling for research on mental health care (Cheng, 2006, Bender, 2005, Kostelny and Wessells, 2005, Chattergee, 2005, de Jong et al., 2005) and

Figure 2-4: Sequencing of emergency and recovery effort (schematic)

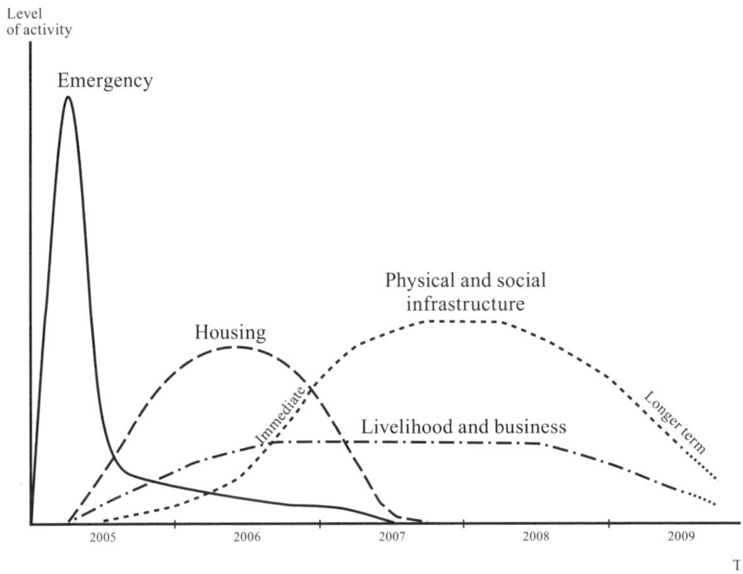


Source: BRR and International Partners, December 2005

public health leadership (Quinn, 2006, Nates and Moyer, 2005, Atkins and Moy, 2005).

Research will help societies to prepare for future natural disasters and limit their secondary effects. For example, in the US research on community reconstruction after Hurricane Floyd in 1999 was published as an employer's manual in immediate response to Hurricane Katrina (Weyerhaeuser, 2005). Further studies are required on how to help communities rebuild after an emergency period. They should include analysis of the formation of self-help groups, the provision of home care for isolated and vulnerable people and the support of income-generation projects (HAIa, 2005).

Figure 2.4 is a schematic of the types of recovery effort following a tsunami; it illustrates how recovery efforts change over time. It indicates that housing needs may remain unsolved longer than it suggests. For example, although the temporary shelters built after Hanshin earthquake were meant to stand for just one year, it took three years to start closing them down and five years before the last was closed down. Although wealthier people were able to find a next place to live in one year, the poor remained in temporary shelters until they received public reconstruction housing or moved, often reluctantly, to somewhere else.

## Conclusion

A survey of the literature on natural disasters showed that there was a major gap in terms of analyzing the long term effects of disasters. The research that was available was mainly focused on the consequences of disasters on mental health, rather than looking at survivors in their social context. The Great Hanshin Earthquake is a unique example because of the number of homes that were destroyed in a major urban area and the extent of the re-housing efforts. Although it is common for a society's most disadvantaged groups to suffer most in a natural disaster, the earthquake was also unique because a high proportion of people killed or made homeless by the disaster were elderly. The great majority of survivors who were still in temporary shelters, or only newly re-housed, five years after the disaster were elderly women.

Although the great majority of elderly survivors were women, reanalysis of the Hyogo District surveys showed that the authorities had not considered gender as an important survey variable. There was a gap in the research to fill by focusing on the problems of elderly people and by including gender in the theoretical framework. The other main research gap identified was the role of the media in social policy making. Campbell (1992) had analyzed the role of the Japanese media in policy change, but very little had been written about the influence of the media on mainstream social policy making. By the final stages of analysis, Klinenberg had published his work on the Chicago heat wave and it was possible to replicate some of his findings on the way the media responded to disaster, as well as highlighting the cultural differences between Chicago and Kobe.

Key texts by Morrow (1999), Taylor (2003), Takahashi (1997) and Klinenberg (2002) set the framework for observation and analysis in temporary shelters and public reconstruction housing. Whereas Morrow and other writers stress the importance of family for survivors of natural disasters, the key issue in this fieldwork was the absence of family. Social isolation (as discussed by Klinenberg) and the Japanese notion of *kodukushi* were identified as important themes for understanding media coverage as well as the efforts of community development workers.

The community development literature revealed a tension which was important in understanding the post-earthquake reconstruction. Western authors, taking Taylor (2003) as a key text, were in favor of bottom up approaches. In contrast the Japanese tradition of community involvement was both top down and male led, though the gendered distribution of power was taken for granted rather than as a problem for analysis. However, the Hanshin earthquake presented new challenges for the traditional Japanese model. Although the literature mostly analyzed the age and sex composition of communities, it generally did not take into account the income of residents, nor did it explain why, five years on, the population of TSH and PRH was heavily skewed toward people without families. These people were predominantly women with low income who were past pension age. The challenge of community building was intensified by the necessity of housing survivors in high-rise flats because of the limited land areas available for development. Subsequently, neither the Japanese traditional model of community activity, nor general models for community development are adequate for analyzing the Hanshin earthquake reconstruction, but they were useful in highlighting characteristics of post-earthquake efforts at community building among the survivors. Attention to gender is essential.

Although the Japanese literature uses many of the same terms as the Western literature, the meanings are contextual and culturally based and so do not translate directly to the Japanese context. The review also identified social differences in Japan as compared with other cultures. These differences include the low prevalence of violent crime in Japanese society compared to the United States and the lack of visible race and class distinctions in the social structure in comparison with both the United States and the United Kingdom.

This literature review was mainly based on published articles on the immediate medical and health impacts of the 1995 Great Hanshin Earthquake and the mid to long term effects of natural disasters in 1990s. Taken together, this contributes to knowledge about the long-term effects of the recent large-scale natural disasters such as the 2004 Niigata Chuetsu Earthquake, the 2004 Indian Ocean tsunami, the 2005 hurricanes Katrina and Rita and the 2005 Fukuoka Earthquake.

# *Chapter Three*
# *Methodology*

Chapter one showed how elderly people have been a focus of Japanese policy making, due not only to population ageing but also to historical and cultural aspects of the health and welfare system. Chapter two described the short-term effects of the 1995 Hanshin earthquake on elderly people in Japanese society, and its implications of Japan's ageing society. Elderly people comprise an increasingly important group in Japanese society; this research is a model for future research on the ageing society in Japan.

Chapter three describes the methodology and setting of this research, including the description of the data and critical evaluation of sources, including the Hyogo Public Health Surveys, media data and field notes of observations and interviews. It will also discuss problems with the analysis of each kind of data. Finally, it will reflect on the application of English qualitative analysis programs like NVivo in a Japanese context.

## Methods

Using qualitative and quantitative methods in a flexible design (Robson, 1993, Cresswell, 1994 and 2003), this research was exploratory; it started by looking at the lives of elderly earthquake survivors in TSH (temporary shelter housing) and PRH (public reconstruction housing) and sought lessons for the future ageing Japanese society. It aimed to uncover a set of case studies of the reconstruction process in the communities. These case studies were focused on individual stories, such as the cases of Mr Yamada (see chapter seven), Mrs Sakura and Mr Tanaka (see chapter six) and the stores of whole TSH and PRH communities. Some individual case studies, such as those of Mrs Sakura and Mr Tanaka, were designed to explore leadership in the communities. Others, such as the study of Mr Yamada, provided a discourse on the experience of loneliness in the communities. The case studies of TSH and PRH communities were aimed at understanding aspects of community generation.

This study looked at low income people and did not represent the full scope of Japan's ageing population. Nonetheless, an understanding of the elderly poor can shed light on Japan's future. PRH was not of the same standard as the wood built pre-earthquake houses that were traditionally the homes of low-income households. Rather, it was comprised of pleasant, modern apartment buildings. These buildings serve as an example of how elderly Japanese people will be housed in the future. The research design followed the methods set out by Robson (1993 and 2002).

Although the research mainly drew on qualitative data, it also used quantitative approaches. For example, it used data from the public health survey and media data. The public health surveys were carried out in the earthquake-affected area over a period of three years by the Department of Health. These data were used as secondary analysis to provide background information for the case studies of individuals in TSR and PRH. They illustrated the way that government saw the aftermath of the earthquake and the needs of the survivors by what information was collected and which definitions were used (for example, the very weak attention to gender differences), as well as the way that changes in this set of values were related to shifts in the values reflected in media reporting. In other words, the surveys reflected government assumptions about survivors and their problems. The reports make it clear that the earthquake left long term problems for many in the older age groups and that their health and social needs were high.

In summary, this study is based on the analysis of qualitative and quantitative content from media reports and on case studies using multiple methods, including ethnography and participant observation, formal and informal interviews and the analysis of documents and records, such as reports by the Hyogo Prefecture Nursing College (1996) and the Department of Health of the West-ward of Kobe City (1998).

## Data sources

### *Media data*

In Japanese society, the media is extremely influential. Campbell (1996: 187–212) showed the important role the media took in formulating the free medical care policy for elderly people in 1980s. After the earthquake, the media focused on elderly survivors and this drew the attention of this research.

Media data was collected from TV, newspapers and fieldwork interviews, mainly over around ten days during the fourth and fifth anniversaries of the earthquake (January 1999 and 2000), in addition to the published reports and books produced by the media. Following Altheide (1996: 19), this data was grouped according to time 'waves' from late 1998 to January 1999 and from late 1999 to January 2000. This intensive method of sampling helped to reveal important thematic patterns that might have been lost had long-term random or stratified sampling methods been used (Altheide, 1996: 21).

TV coverage about the earthquake-affected area was recorded and transcripts were prepared for eight days each year. The lengths of videotapes were 27 hours 55 minutes in 1999 and 38 hours 55 minutes in 2000. Two other special feature TV programs, which aired on 24 November 1998 and 13 June 1999, were also analyzed. While transcribing the TV programs, the narrator's voice was translated into English. Visuals were also described in English. The transcriptions were helpful in enabling closer examination of content and for developing a framework for dealing with visuals in the TV data analysis (Altheide, 1996: 20). TV data was mainly from news programs, special features and documentaries on earthquake related topics. Tables A3.5.1 and A3.5.2 show the framework of the 1999 wave and the 2000 wave respectively. Newspaper articles normally have headline news on the first page, local news in the back pages, and editorials.

Transcriptions were made of all the media data but gradually the focus shifted to coverage of elderly people, housing and health. They were sourced from TV stations on air in the Kansai (Kobe and Osaka) area, the national station NHK (*Nihon Hōsō Kyōkai*, Japan Broadcasting Corporation, a public service broadcasting organization like the British Broadcasting Corporation), other privately owned stations at the national level (Asahi, Mainichi, Yomiuri, Kansai-TV) and local (Sun-TV, TV-Osaka) levels. This media data was complemented by interviews with a TV reporter and a newspaper journalist. The interview with the TV reporter was transcribed.

Media data was collected in parallel with data collected in the field. This allowed for the analysis of the media's discourses alongside my own discourses. Some of the people and places in the TV coverage overlap with the people met or places I have visited for my fieldwork. The fieldwork data collection will be explained later in this chapter.

Newspaper articles were collected from *Asashi Shinbun* and the *Kobe Shinbun* and translated into English. *Asashi Shinbun* is the most widely read newspaper in Japan. It had a circulation of more than 12.6 million readers in 1995[1] (Cooper-Chen, 1997: 52–53). Articles were also collected from the Osaka edition of the newspaper, including from both the front and back pages, the latter dealing specifically with local news. The Osaka edition covers the earthquake-affected area of Osaka and Kobe. The other newspaper, *Kobe Shinbun*, is a local newspaper. It is the major newspaper of the locality and therefore has direct contact with the issues arising in the area.

A3.4 is a list of *Asashi* newspaper articles on post-earthquake topics from the January 2000 wave. Others, including those from 1999, were not translated into English except for sections cited and used in the following chapters. They were not all translated into English and entered into NVivo because most articles did not add to the data already collected. Many contained only figures for *kodokushi* (dying alone) or the number of residents in TSH and PRH. Japanese newspaper headlines are often comprised of incomplete sentences; they are expressed in shortened phrases with key words. This can be seen in some of the problems with English expression in the Appendix. Any attempt to make sense of the headlines is a matter of analysis rather than a direct translation from Japanese. While translating into English, I was aware of the problem and as a result the records of newspaper based data are not as extensive as the TV lists. Also, while newspapers are issued only twice a day, TV programming runs all day, from morning to midnight. I decided it was more important to focus on fieldwork and TV data while not ignoring newspapers. Only the *Asashi* evening editorials articles were analyzed in NVivo.

The media tends to focus on visually exciting images and events (Altheide, 1996: 9, Klinenberg, 2003: 217) and this might have been the case with the media coverage of the earthquake survivors. By the time of the research, the TSH were near to closing down and the remaining residents often had problems that were extreme and not typical of the general TSH population. As a group, they were no longer representative of Japan's ageing population, but they could be seen as an increasingly important group who had been otherwise invisible. Media coverage of those in PRH was less extreme in comparison to the fieldwork reports. These were people whose move from TSH meant the end of emergency living and the beginning of post-disaster normality. In terms of life reconstruction they were progressing. They had decent housing, solving an important problem in the earthquake recovery process.

### *Hyogo Prefecture Health Surveys, 1996–1998*

The Hyogo Prefecture Health Surveys placed data on the research population in a wider context. Three volumes of the annual survey, from 1996 to 1998, were analyzed in order to identify and compare changes over time. The questionnaires varied each year, apparently in response to the findings of the previous years by the Department of Health. As discussed earlier in this chapter, secondary analysis enabled me to examine the Department of Health's point of view, focusing on what they saw as important, how they described the process of life reconstruction and how they measured the process, including their community interventions.

Table 3.1 shows the response to the Hyogo Health Survey from 1996 to 1998 by housing categories, illustrating the changes in the surveyed housing cat-

Table 3-1: 1996–1998 Hyogo Health Survey response rate

| | 1996 | | 1997 | | 1998 | |
|---|---|---|---|---|---|---|
| | Number | Response rate | Number | Response rate | Number | Response rate |
| Temporary Shelter: responses | 5,315 | 57.1% | 3,644 | 56.5% | 817 | 46.2% |
| Number distributed | 9,315 | | 6,451 | | 1,769 | |
| Public Reconstruction Housing: responses | | | 3,165 | 70.9% | 6,248 | 56.9% |
| Number distributed | | | 4,465 | | 10,982 | |
| General: responses | 902 | 33.2% | 1,029 | 27.3% | | |
| Number distributed | 2,714 | | 3,772 | | | |

Source: Hyogo Prefecture Department of Health (PRH and G), 1996-1998.

egories over time. The population of the first survey in 1996 was comprised of residents of TSH (temporary shelter housing) and general housing only, because PRH (public reconstruction housing) had not yet been built. The population for the 1997 survey was residents of all the three housing types, including PRH. By 1998, residents in general housing were no longer included.

The disadvantages of postal and other self-administered surveys were evident in the data. There may have been ambiguities in, and misunderstandings of, the survey questions that were not detected. On the other hand, interviewing would have been too expensive for such large samples and a postal survey is the easiest way of retrieving information about a large set of people. They can be extremely efficient in providing large amounts of data at relatively low costs in a short period of time (Robson, 2002: 233). They are therefore suitable for gathering information from people moving into PRH and from dispersed populations in general housing. They also allow for anonymity.

In interviews, the interviewer can clarify questions and the presence of the interviewer encourages participation and involvement. This was even more so the case for public health nurses who sometimes knew the respondents. They could often help them to answer the questions more accurately. On the other hand, nurses might bias results by filling in answers or even whole forms. The nurses might then offer their own views rather than those of the elderly people being interviewed.

In order to be valid, the response rate has to be at least 70%. Some say the response rate has to be 90% if biased estimates are to be avoided (Robson, 2002: 251). The Hyogo Public Health Surveys did not meet the 70% response rate (Table 3.1). The sampling and collection of the questionnaire responses was different depending on the housing type. Because of this and the other usual problems with large scale health surveys (Robson, 2002), the results may have been biased. The situation of the sampled groups could be better or worse than indicated by the survey, but it is impossible to be sure.

The response rate from respondents in TSH was lower than that of previous years at 46.2%. There are a number of possible reasons for the rate being so low. First, in 1998, official public health nurses that had in previous years visited TSH everyday only visited once a week at most since there were no longer many TSH residents. Second, it was also apparent that many of these nurses did not have rapport with the TSH residents. They were even asking volunteer public health nurses for instruction as they thought volunteers knew better. Third, a larger proportion of the TSH residents in 1998 were what the public administration called 'problem cases' who had have given up hope and had no motivation to fill in the questionnaire. The low response rate may have biased the results and the general situation might have been worse than the survey results showed.

The highest response rate was from the PRH residents in 1997. It was the only group which responded at a rate of over 70%. This group had more hope than other groups and the questionnaire was an opportunity to express their opinions. Despite the fact that public health nurses had more rapport with the residents in TSH than in PRH, the response rate was higher in PRH.

The questionnaire design changed over the years. In 1996, the first year, the survey served as a pilot to help design the questionnaires for the 1997 survey. The questionnaire was further revised in 1998 in response to the findings of the previous year, especially in relation to the issues that respondents raised in open-ended questions.

The survey sample of elderly people was more representative than that covered by the media. The TV was mainly concerned with disadvantaged elderly residents of TSH and PRH. The first two years of the surveys collected data from general housing as well but the sample size and response rate were much smaller. The data on general housing was collected to provide a comparison with the residents in TSH and PRH, the main concern of the survey. This might indicate that the attention of public administration was focused on the residents at TSH and PRH and that they expected people living in general housing to be self sufficient. Hirayama (2000) pointed out that the post-earthquake housing recovery policy considered many victims as self sufficient, not needing temporary housing under the Disaster Relief Act. Since being a resident of TSH was one of the categories that gave residents priority in most PRH allocation processes, those who were excluded lost their chance of winning an apartment in PRH. These people disappeared into general housing and it became more difficult to reach them in surveys. A large proportion of people in general housing might have been in middle-income households, but the survey sample cannot confirm this as it does not show the range of residents in general housing and the survey stopped collecting data from 1998. The general housing population could be more representative of the future Japanese population, but the population groups living at TSH and PRH remain important groups needing public attention. People at TSH and PRH were poorer and more deprived than the average; they were more severe cases.

The findings from the secondary analysis of the survey are presented in chapter four.

### *Interviews*

I interviewed four public officials at the Department of Health of Kobe City on 26 January 1999, four officials at the Department of Health of the West Ward of Kobe City from February to August 1999 and one official of the Central Ward of Kobe City on 25 February 1999 (See A3.1). The Director of the Department of Health of Kobe City was introduced to me by the then Director of Planning Division of Welfare for the Elderly of the Japanese Ministry of Health and Welfare in Tokyo. He brought other officials from his department to our one hour meeting on 26 January 1999. He explained the work that his department did for earthquake survivors, mainly in area of public reconstruction housing, and his views about the state of the reconstruction process and future priorities.

Per the director's request, his staff introduced me to the two PRH sites, Natsu Aki PRH. Fuyuyama PRH (a main fieldwork site) and another PRH site where I visited only once. These first two PRH sites were described as successful cases with active community leaders. The officials explained that there were 2,300 Silver Housing dwellings in total, with two PRHs having Silver Housing in the Chuo District (the Central Ward) and the Fuyuyama PRH. The other PRH was visited only once because there had been just one dinner meeting and no other community activity had started. The officials thought that it was not worth visiting. Three other officials of the West Ward and the Central Ward of Kobe City were visited and their community activities were observed.

For the purposes of this research, the selection of the field sites would not have been different whether they had been chosen by the Department of Health and Welfare of Kobe City or not. The sites that were chosen had started community activity at the time of the research while others had not.

The official introduction made it easier to access research sites, but it did not introduce bias in terms of the site selection. Nurses with the Department of Health and the chief of the self-governing body of Fuyuyama PRH were very cooperative. Public health nurses would have been less helpful in providing information or being available to explain the background if I had not been introduced by their supervisors. A small bias could have arisen because the chief of the self-governing body, Mr Tanaka, wanted to make a good impression, but he was open in his approach and let me talk with anyone, including those who criticized his leadership style. LSAs (Life Support Advisers for elderly people) were expecting to be interviewed by a visitor like me and were eager to explain what they were doing. Interviews were unstructured and detailed notes were taken during the interviews.

### *Ethnography*

Three community locations were selected for ethnographic study by theoretical sampling (Table 3.2). Works on theoretical sampling were consulted (Arber, 1993, Flick, 1998, Robson, 1993 and 2002). My research was exploratory and employed theoretical sampling, rather than probability sampling (Arber, 1993: 72). In theoretical sampling, individuals and groups are selected according to their expected ability to provide new insights for the developing theory (Flick, 1998: 65). Location, conditions of facilities, types of actors and activities were all considered in the selection process, as were the different types of housing schemes such as individual apartments, shared housing, and Silver Housing. This would illustrate whether or not community generation differs by housing and under what conditions might loneliness can be alleviated.

Participant observation was done within TSH communities. Direct observation tells stories which do not otherwise come out of surveys and media data. Four different temporary shelter communities were visited and one temporary shelter community in suburban west Kobe was selected for participant observation. Two were almost about to be closed and no longer had community activities. A *minsei iin* (welfare commissioner) who used to work there regularly only came to the TSH to assist with my visit. The other TSH was difficult to visit regularly because of the reason explained in the next paragraph. The site I selected was the second Haruyama TSH. The Haruyama TSH had three parts with the second part being the largest with 250 household units. I worked as a volunteer waitress at a weekly tea shop at the community center from November 1998 to May 1999.

The second Haruyama TSH was the easiest to get into for fieldwork. The head volunteer public health nurse was open-minded and said she believed

Table 3-2: Three fieldwork sites

| | TSH: Haruyama | PRH 1: Natsu Aki | PRH 2: Fuyuyama |
|---|---|---|---|
| Location | Suburban | Suburban | Urban |
| Kobe city | West Ward | West Ward | Central Ward |
| Building type | Hutted apartments (*nagaya*) | High-rise | High-rise |
| Number of household units | 250 | 700 (11 buildings) at Natsuyama PRH and 500 (3 buildings) at Akiyama PRH | 550 (2 buildings) |
| Silver Housing | | | 200 (lower levels of one building) |
| Leader | Volunteer nurse | A woman welfare commissioner, chief of the self-governing body and of volunteer neighbourhood community | A male chief of the self-governing body |
| Methods | Participant observation | Shadow a community leader | Observation at events |

anyone could do volunteer work, that there must be something each person could do. On the other hand, at the more famous TSH that often appeared in media coverage, the volunteer leader said that amateurs could not handle this very difficult situation and would have to be trained before they could participate as volunteers. At this second TSH, I was able to see the situation of elderly people left behind at a TSH after the peak period. I saw some nursing college and graduate students working as volunteers. As the leader took the responsibility very seriously, she lived at the TSH. Under her management, the team atmosphere at the volunteer's office was strict. It appeared that you had to show a serious commitment to be part of the team. This TSH would have added a variation in leadership style to the research—the style of *renkei* (coordination) public administration coordination. But at the start of fieldwork, community was disappearing and conducting a fieldwork at this TSH would not have given much new data.

At the Haruyama TSH, I accompanied a public health nurse patrol on home visits in the community on several occasions from March to April 1998 and from August 1998 to February 1999. I selected this TSH for several reasons.

First, this was the place I had visited in the very beginning of the research project. Second, activities such as the weekly tea shop at the community center were still held, whereas other TSH communities had started to close down such activities as people moved out. Third, it was easy for me to go into the community because of the rapport I had established with the volunteer public health nurse and the shelter housing residents. I was able to work there as a volunteer sponsored by her NGO and they paid my local transportation fee for the activity. Fourth, this shelter was established within the same municipal district as one of the public reconstruction estates that I had selected for my fieldwork. Fifth, this was the only shelter community where the Department of Health had prepared a report which I could use for secondary data analysis. Sixth, fieldwork at the Haruyama TSH enabled me to follow the individual case study of Mr Yamada who moved from the Haruayam TSH to the Natsuyama PRH.

I carried out fieldwork at two PRH communities and made short observation visits at three others. I visited the sites from February to July 1999. The two PRH sites selected were chosen for me by the Department of Health of Kobe City. They chose the PRH sites where they believed the research would be most successful. One was where the *minsei iin* was active. The other was where they said the chief of *Jichikaichō* (the self-governing body) was active. As two PRH compounds with Silver Housing had been built to respond to the high proportion of elderly people needing post-earthquake public housing, I requested from the Department of Health access to both. I visited the other place only once in 1999 and was asked to join their dinner, which was the second of their monthly dinners. I visited the other two PRH sites in 1999 when I joined an NGO survey team visiting each PRH household to discuss employment opportunities as part of life reconstruction. These places seemed to have no community activities occurring. I would say the community was not dynamic and I did not expect to see much change.

One of the main PRH field sites visited was the Natsu Aki PRH, where *minsei iin* were active. I visited there five days in February and four in July 1999, observing both morning and afternoon activities for most days. This huge community could be divided into two major groups: eleven high rise buildings of 700 household units in total with a high proportion (70%) of elderly people living alone, and three high rise buildings of 500 household units for family apartments.

A number of reports prepared in 1996 by Kobe City West Ward health officials provided the statistical background on community generation for ethnographic fieldwork in TSH sites.[1] They were useful tools for measuring how representative the informants were of the overall population. It also verified my field observations, as well as the *gyōsei*'s views on the community situation and the work done in the community. It also provided information about relevant NGOs, which ones were still active and which ones were not.

The other main PRH research site was the Fuyuyama PRH, in central Kobe, where it was said that the chief of the self-governing body was active. I visited there three days in February 1999. This community was smaller than the first one, although the residents felt that they had moved to a big building scheme. There were two high-rise buildings. The number of households was 550 household units, 220 of which were Silver Housing. It had one of only two Silver Housing schemes in Kobe City Central Ward built as post-earthquake public housing.

This case study showed the potential achievements and some of the problems faced by a charismatic community leader. The personality of Mrs Sakura was special; it shows what can be done in such circumstances. A longer study would be needed to show how far the experience of Natsu Aki PRH could be transferred to other sites and how far individual characteristics were definitive and essential.

The environmental setting may affect how well community generation projects run. The selected fieldwork sites offer the chance to examine the process of community regeneration following the earthquake. Choosing sites with different characteristics permitted the analysis of the variations in the processes of community regeneration, such as differences in leadership and housing aspects.

The research was limited in that the PRH field site selection was biased by the perceptions of those who introduced me to the sites, such as by the Department of Health of Kobe City. But from my visit to other inactive PRH sites and from what I heard from advisors working at Natsu Aki, it was the best available sampling method.

Community development in high-rise PRH apartments and single story TSH encampments will be compared in chapter five.

## Characteristics of the researcher

This section will discuss how various characteristics of the researcher influenced the study, particularly her gender. Gender is a crucial aspect of ethnographic research (Arber and Ginn, 1995); it affects the capacity to access, move about in and observe public spaces and women's perceptions of limitations to access are much more sensitive than men's. That makes a woman researcher observe differently and notice things differently compared to male researchers (Flick, 1998). Warren and Hackney (2000) examine some of the issues for women working in the field. Their work informs this discussion.

### *Entering the field*

In the TSH site, I was accepted as a volunteer, in the PRH site as a research graduate student. I am from the Osaka-Kobe area, but the area studied was not part of my own neighborhood. I did volunteer work at the Second Haruyama TSH from March to April 1998 and from November 1998 to August 1999. I accompanied a volunteer public health nurse on her home visits. She had already gained the trust of the community after working there for three years. It was not always easy to gain the residents' trust, but the nurse's welcoming words encouraged me to continue the work. I also worked as a waitress at the weekly tea shop along with the nurse and her Christian friends.

My being much younger than the volunteers influenced the research in a way that I did not realize in the beginning. Elderly people who would otherwise only meet with other elderly people were happy to meet with a younger person like me. Some older people complained about their daughter-in-law and said that they wished that I was their daughter-in-law instead. Some older women said that they wanted me to serve the coffee rather than other older volunteer waitresses. There were two twenty-year-old nursing students who also worked as volunteers. They were very nice, polite and they worked hard at cleaning and other tasks; they wanted to do something for the people at TSH, but not just sit with them and listen. But perhaps I was better at listening to the older women.

I did not actively participate in the activities at the Natsu Aki PRH. Government contacts, from the central government in Tokyo to the Department of Health of Kobe City, introduced me to *minsei iin* in the West Ward of Kobe City. They were the gatekeepers to the Natsu Aki PRH. They were taking the initiative in various activities, leading the community regeneration projects that will be discussed in chapter six.

### *Gender*

The young women who were typically employed in the PRH health promotion program were either fixed-term staff of the city specially employed for the post-earthquake recovery or visiting nursing students. At other community activities, young women were typically either employed by the city or were volunteers. At the TSH sites, young women were typically either visiting nursing or social work students or volunteers. I was generally treated as one or another of these social types, though once a taxi driver asked me if I was a journalist. On the other hand, conversations with other visitors revealed that male visitors tended to be taken for medical students, doctors or permanent staff of the municipal government.

At breakfast in the Catholic Church dining hall at the Fuyuyama PRH, I was treated as if I was staff from the municipal local government, even by people to when I had been introduced to more than once by the chief of the self-governing body as a researcher from the University of London. On the other hand, male visitors would be taken for university professors. I had expected the elderly residents to see me as a volunteer or a public health nurse, but I was sitting next to a fixed-term staff member of the municipal government so they thought that my position was the same. Also, they must have been visited by many professional women employed by the Department of Health, such as public health nurses and nutritionists.

On one occasion, while participating in the health promotion program at the Fuyuyama PRH, an elderly woman thought I was a nutritionist or public health nurse running the program together with the Department of Health. These tend to be considered women's occupations. I had only intended to observe the activities but I eventually participated because the elderly women talked to me and asked me to do things like open a sauce bag with my fingers. It was not just the case that I did not know them, but they did not know each other because it was still the early days of the program. Most were too shy to socialize and it looked as though I could facilitate this. I was able to do little things such as encouraging them to start talking to each other by listening to what they said. This also helped me with my research interests as I learned about their daily patterns without spending many more hours observing them. In this way, my being a woman might have had advantages for the research project. Public health nurses, who are generally women, have a long history in Japanese health at the field level. I might have sometimes been taken as a public health nurse. It is possible that the residents at TSH and PRH accepted me as a woman as they were used to having female nurses around.

These traditional ideas about gender might be changing with new generations. In the younger age groups, more men are now going into areas that were once thought of as women's work and vice versa. This might be due to the influence of other cultures, or the adoption of gender-free titles, although Japanese language does not gender words to the extent that English does. It might be due to international pressure to recognize the legal equality of the genders. It might also be because it is more difficult for new graduates to obtain permanent life-long employment, a once common practice in Japan that is now challenged by economic stagnation. But the generation of interest in this research understands that the role of public health nurse is one normally taken by a woman.

As a woman, it was easier for me to talk to the elderly people because most of them were women themselves. Elderly women participated actively in various community activities such as clubs and community meetings. However, it was rare for elderly women to be office holders and when they were they were often treated disparagingly. For example, I observed that when women argue at meetings, men will often say, 'You are just a woman, shut up' or, 'Why is the district sending a female representative?'. This information was shared so openly because I am a woman.

Mr Tanaka, chief of the self-governing body at the Fuyuyama PRH would call me '*Sensei*', which is a title for teacher, doctor or professor. As I am a dentist, I have been called '*Sensei*' by my patients. However, I was not there as a dentist, but as a graduate research student introduced by the Department of Health of Kobe City. Nonetheless, Mr Tanaka and the LSAs thought I was a professor, probably because I was called '*Sensei*', or because Mr Tanaka was happy that his work received professional attention and he was proud to introduce someone as such to the people he worked with. The chief of the public health nurses unit in the ward of Fuyuyama PRH would call me '*Sensei*' because I am a dentist.

Mr Tanaka did not consider my gender when he located me in the social hierarchy. He was more interested in my educational background, that it sounded academic and professional. I did not ask, but it was likely that he had not received a university education, since he was living in public housing. He used some words to distinguish himself from other members of the resident's community. For example, he felt he should 'train' members to be able to do this and that in the community. His approach was not democratic or bottom-up, but based on the top-down leadership of one man. At breakfast in the neighboring Catholic Church dining hall when Mr Tanaka had left for a dental appointment, members of the self-governing committee and other residents said, 'This way of talking can be accepted because he is a man. If a woman spoke this way, she would be hammered down'. At least on the surface, people might prefer to have a male as a leader as they are more comfortable with men being assertive and taking on leadership roles.

The third PRH with Silver Housing I visited had a male chief of the self-governing body in order to make things run smoothly. But most people believed that it was the female vice-chief that made most of the decisions. The female vice-chief herself often emphasized that it was the chief who made the decisions so that she did not appear to be assertive herself and express her opinions in public. She did not wish to be chief officially herself.

As a female worker, I felt that I was seen by both women and men as less-threatening and more open to emotional communication than if I were a man. My being a woman probably made it easier for some men to talk openly, especially about the inner worlds of their feelings and thoughts. This was an advantage, as a man in TSH might not have told the same story to another man. For example, a man in TSH who had leukemia was very clear about whom he would socialize with. He kept his distance from other TSH residents but wanted to make friends with outsiders such as the volunteers. He felt that his illness prevented him from going into a crowd, but it was possible that there was a gender aspect to his reluctance to socialize with other residents in the tea lounge, most of whom were women. He preferred to talk to volunteer waitresses because he was afraid he did not have anything in common with the other men and if he were seen to be talking to other women, it could lead to gossip. With the volunteers he did not have this problem because they were from outside the community.

In another interview during my fieldwork, conducted in April 1998, the head of an old people's home said that managing the gender balance was a challenge. Even though only a few residents were men, the men woould not always become friends with each other.

### *Language*

My native language is Japanese from North Osaka, close to Kobe and Kyoto. Having worked in Tokyo and abroad, I have met many Japanese who speak with a Tokyo accent and I can talk with a Tokyo accent. Some Osaka people ask me why I speak with a Tokyo accent when I am from Osaka. I am proficient in the Osaka-Kobe dialects so I had few serious problems with using language in the ethnographic context, although I encountered some language differences among different age groups. Also, I have a very close relationship with my grandmother from Kobe, so I have no problem in understanding Kobe accents.

After the earthquake, there were new words created and used frequently in the Kobe-Osaka area that researchers from Tokyo did not understand. The analysis of these words can help to illustrate aspects of the post-earthquake reconstruction process and the perceptions held by various people and the public.

The research was undertaken under the supervision of a Western woman. This influenced the development of the language and tone of the research findings. For example, the research uses concepts like gender, cultural particularity and sensitivity. She helped me to be more reflexive when talking about Japanese culture by asking me to explain and elaborate on my assumptions.

English words differentiate between genders more than Japanese, but Japanese colloquial expressions are more gender specific.

### *Dress*

When interviewing public officials and journalists in an office, I dressed in a suit. At the TSH, I dressed very casually, wearing inexpensive clothes such as a sweater, a pair of pants and sneakers. I followed the dress style of Mrs Ogawa, a public health nurse volunteer. I noticed other volunteers among her Christian friends dressed much more nicely. Mrs Ogawa, had won the trust of the TSH residents who did not dress very well. Because I imitated her style of dress, the residents were friendly and open from the beginning of my fieldwork.

At Natsu Aki, I did not dress in a suit so as not to look like a career woman. Like Mrs Sakura, I often dressed in a nice but casual one-piece dress in summer and a warm, casual suit and sweater in winter. I dressed more nicely than the residents, more like a welfare commissioner or housewife volunteer from the neighboring east-side community, or like a graduate nursing student. On one occasion a resident asked me if I was a journalist.

### *Appearance*

Although differences in skin color are less significant in Japan than elsewhere, some skin tones are often associated with the working class. Outdoor labor can damage skin, making it look older and darker. However, this association is more often applied to men than women.

### *A Japanese researcher at a Western university*

Although I was registered in Osaka and went to school in both Osaka and Kobe, my current affiliation with the University of London and my years spent living abroad might have made me an outsider. The NGO leader who obtained the Hyogo Prefecture Health Survey data for me said that he had explained to the prefecture that it was to be used by a foreign researcher.

## Data analysis

Both quantitative and qualitative methods were used to analyze the media data. Discourse analysis, as outlined by Altheide (1996), comprised the main qualitative methodology. Case studies and discourse analysis were used to examine field notes. Literature on methodology (Robson, 1993, Yin, 1994) and interpretation were consulted (Strauss and Corbin, 1990 and 1998, Antaki, 1988, Bryman and Burgess, 1994, Gubrium, 1988, Silverman, 1994, Richards, 2005). Public health surveys were subjected to secondary analysis. Thomas et al. (1998) is an example of literature in this field. Narratives were used preliminarily to organize the data and then they were subjected to discourse analyses with and without computer assistance. A large body of literature on discourse analysis was consulted (Gill, 1996 and 2000, Potter, 1996, Potter and Wetherell, 1987, Wetherell and Potter, 1988, Wetherell et al., 2001, Kleinman, 1995, Burman and Parkereds, 1993, Bauer and Gaskelleds, 2000, di Gregorio, 1999, Fraser, 1999, Richards and Richards, 1994, Richards 1999a, 1999b and 1999c, Bazeley and Richards, 2000, Robson, 2002: 464–472). According to discourse analysis, explanations for the same phenomena vary depending on the reasoning employed by different people, groups and authorities.

Social scientists in Japan are taking a growing interest in qualitative research (Suzuki, 2000), but there are no books in Japanese that describe qualitative social research in the social sciences other than recently translated American books. Emerson (1995) was translated into Japanese in 1998, Vaughn (1996) in 1999 and Strauss and Corbin (1998) in 2004. The translators of Emerson's book wrote in the afterword that there were a number of words and issues could not be easily translated for the Japanese context. They explained that they exchanged e-mails with Professor Emerson for further clarification of those issues (Sato et al., 1998: 492).

The quantitative analysis of media data is described below, while the qualitative analysis is incorporated into later chapters. The media produced a range of discourses on the earthquake; they changed over time in a way that was parallel with the relocation of survivors into new housing. This process of change included volunteers, different NGOs, the welfare commissioner, public health nurses, public administrators, the media and residents in Kobe and surrounding areas. All of these individual and group agencies had a key role in community generation after the earthquake.

### *Media data analysis*

The data was sorted by TV station type, i.e. public and private, local and national. Each station had its own mandate and that there were biases in each presentation. These changes in tone were analyzed over time, particularly in so far as the changes concerned gender issues. Gender analysis was of particular importance because, while it was an important aspect of the research project, it had been lacking in the original analysis of the public health survey. Interviews were conducted, transcribed and analyzed to see what was behind the publicity. The analysis of media reporting raised questions which helped to focus the fieldwork.

Media analyses revealed how the media focus shifted over time, and how these shifts reflected the issues that mattered to Japanese society. In particular, the media analysis drew attention to the plight of elderly people living alone. Through interpreting the media discourse, it was clear that an important concern was that of the problems of elderly people in the context of Japan's ageing society.

The process of transcribing the data into English helped to reveal some of my own assumptions which I might otherwise have taken for granted as a Japanese person. Had I not undergone this process, I might have missed some points that were noteworthy.

It is important to question the validity of media data as it can be biased. Interviews with a TV reporter and a newspaper journalist at the Kobe Shinbun confirmed the existence of background politics within the TV station. Differences between TV stations and newspapers often reflect such biases. While this is an important point to remember, my research was not about politics in media organizations and I did not conduct extensive interviews about this matter with key informants.

Quantitative analysis of media data showed the actual length of the time spent on specific topics. Content analysis (Robson, 2002: 352–359) showed how this focus shifted over the years. It showed how media attention shifted to and away from issues that elderly people face. Content analysis can be extremely laborious and time-consuming. It is a field where computerization has led to substantial benefits (Robson, 2002: 357). The text can be easily manipulated in various ways, such as showing all sentences containing a particular word or phrase.

This research employed the key-word-in-context (KWIC) technique described by Robson (2002: 359). This technique provides a list of contexts in which selected key words appear, sensitizing the analyst to the variations in meaning of commonly used words. It also employed word frequency lists, category counts and the combined criteria list, used for joint occurrences of two words or phrases in a sentence or passage (Robson, 2002 : 359). This retrieval and comparison of the words was extended not only to media analysis but to the analysis of fieldwork and the public health survey questionnaire too. This data triangulation uses more than one method of data collection (Robson, 2002: 174).

### *Secondary analysis of Hyogo Public Health Surveys*

I obtained the Hyogo Public Health Surveys in hard copy form with the tables from each year showing. Unlike with raw data, it was not possible without re-entering the data to conduct cross tabulations using an analytical computer software package such as SPSS, which would be possible with the raw data. Secondary analysis was therefore limited, but nonetheless it highlighted several issues.

As the questionnaires were modified every year, time series comparisons were also limited. But the analysis of questionnaire modifications showed how the design changed to reflect the findings from the survey of the previous year. Although the questionnaire design was modified over the years, some cross-year statistical comparisons were still possible even though questions had changed.

### *Analysis of field observations and interviews*

An advantage of observation as a technique is that it is direct (Robson, 2002: 310). Interview and questionnaire responses are notorious for discrepancies between what people say and what they do. A disadvantage is that it is time consuming, not easy and could, in the act of observing, affect the behavior being observed.

At TSH, I was 'the complete participant' (Robson, 2002: 316). Robson warns about the tendency to 'go native' but my observation period was short enough to avoid this. At PRH I was 'the participant as observer', a role I made clear from the start. By participating in activities, I was able to ask members to explain various aspects of what was going on. It was an important step towards getting the trust of key members of the group.

To reduce observational bias, Robson advises that, where possible, a record should be made of observations on the spot, during the event (Robson, 2002: 322). At TSH, at a weekly tea shop, I carried my notebook in my apron pocket and, upon returning to the kitchen after serving tea and having a chat with the residents, took notes immediately. While accompanying the public health nurse, I took notes in between activities such as cleaning and cooking. Nobody seemed to be bothered or even noticed; community health workers take notes so my practice was not seen as strange. At PRH, I was taking notes on the spot. That might have encouraged some people to speak more; most people did not look as if they were worried one way or another. Once, during a house visit, a man became curious and asked me what I was writing, but he just asked me to write something good about Mrs Sakura.

Robson advises the researcher to make a conscious effort to distribute their attention widely and evenly in order to avoid being selective (2002:

324). Because I was given directions from others, selective bias did not come from me. For example, selective biases would have emerged during my TSH visits because the public health nurse who I was shadowing visited extreme cases that she believed would need attention. But because most people still residing in TSH by the time of research had problems, this bias was not extreme. Also, the public health nurse normally tried to show a range of cases when I accompanied her. At TSH tea parties, those who I met tended to be those who may have been more open than *tojikomori* (those who had closed up on themselves). As those who did not come to tea parties were seen on home visits, this bias was minimized.

To help prevent selective memory, Robson advises writing up field notes in a narrative account promptly (Robson, 2002: 324). After taking notes on the spot, I started to type them into a computer document in the evening, finishing the following day. Typing up field notes took hours, longer than the time actually spent in the field.

Robson suggests 'prolonged involvement' as one way to reduce bias (Robson, 2002: 174). It helps to reduce both reactivity and respondent bias. Researchers who spend a long time in the setting tend to become accepted and initial reactivity reduces. It permits the development of a trusting relationship between the researcher and respondents so the latter are less likely to give biased information. At the TSH, I shadowed Mrs Ogawa who had developed a trusting relationship with the residents and soon began to develop this relationship myself. At the PRH, thanks to the reputation of *minsei iin* Mrs Sakura, I was readily accepted as trustworthy. However, Robson warns that prolonged involvement can also result in greater researcher bias (Robson, 2002: 174). It may be difficult to maintain the researcher role over an extended period of time (the 'going native' threat), or developing antipathy might result in negative bias. Robson advises peer debriefing and support to prevent researcher bias. My supervisor helped me to question whether my good relationship with Mrs Sakura and my introduction to PRH fieldwork sites by the Department of Health and Welfare of Kobe City and the Ministry of Health and Welfare in Tokyo resulted in bias.

Program logic models, a combination of pattern matching and time-series analysis (Yin, 1994:118), present strategies for the exploration of case studies. Pattern matching was incorporated into the analysis of Natsu Aki (Natsuyama and Akiyama) PRH and Fuyuyama PRH, identifying commonalities and variations in, for example, the needs of leadership in community generation. My research supported the theory that local community support is needed for government officials to work in a community and this method showed the variations in local support.

Several points may be made in relation to time-series analysis. It was useful in analyzing the case of Mr Yamada, who changed his attitudes over time as he moved on from the Haruyama TSH to Natsuyama PRH. Complex analysis over time also helped in understanding loneliness, as well as in the variations in time relating to individual housing and community generation in various TSH and PRH settings.

The fieldwork was conducted in the third and fourth years after the earthquake. This was in time to observe the TSH before it was closed down. It was too late to observe the TSH community, but some of the details of community generation could be learned retrospectively. On the other hand, the process of community generation could be observed at several PRH sites, where it was only just beginning. Long term observation is required to carry on the story. Bias can be reduced by a longer term approach (Robson, 2002).

The fieldwork at TSH was carried out from 1998 to early 1999. The first set of media data was collected in January 1999 in the week of the fourth anniversary of the earthquake. The fieldwork at PRH was carried out in the following months and then the second set of media data was collected in January 2000 in the week of the fifth anniversary of the earthquake. This allowed the focus of the research to shift from the media to ethnographic fieldwork and then back to media. It also meant that communities were observed while the media was filming to prepare for the 2000 wave of broadcasting.

NVivo[3] was used to aid in the discourse analysis of media data and content analysis of interviews, field notes, observational data and other relevant documents (Fielding and Lee, 1991). It was also used to identify the different discourses relevant to the aims and objectives of reconstruction, the definitions of outcomes and measures of success and failure, the attitudes towards family and individual welfare, and other Japanese values.

The downside to using specialist software like Nudist is the time and effort needed to become proficient in its use; hands on workshops are helpful (Robson, 2002: 463). NVivo, on the other hand, is much more user-friendly. The transition from Nudist to NVivo was not a big problem. I joined a few qualitative research online discussion groups that were helpful and often encountered timely topics and suggestions, even learning about the hands-on workshops on qualitative research by the developer that I attended in Sydney.

NVivo helped in the analysis of different styles of leadership such as those of Mr Tanaka and Mrs Sakura. It systematically retrieved by category data from individual case studies scattered across field notes from various dates. For example, it helped to follow a single case such as Mr Yamada who had changed from a closed attitude to a more outgoing one over the time of his move from TSH to PRH. It also helped in the search for the names of people who had appeared in both TV news and in my fieldwork, as well as in the search for media discourse on *kodokushi* in the case of total lack of family. NVivo enabled the revisiting of data over and over again while making an argument and the location of evidence in a systematic way, rather than simply eyeballing (Richards and Richards, 1991: 38–72).

### *Application of NVivo to the Japanese context*[4]

I collected media data from both TV and newspapers. As NVivo cannot directly code multimedia data but can only link to it, the TV coverage (67 hours) was videotaped and transcribed into English so that it could be coded as a document. The data was organized according to type and research site. By coding and searching for a person's name in all the documents, variations could be identified and compared. Sometimes the same person appeared in TV video transcripts, newspaper articles and recorded interviews and observations. To handle the data, several document sets were made. For example, the media document set included video transcripts, newspaper articles and interview notes with a TV reporter and a newspaper journalist. This meant various types of data on the media were grouped into one set.

This research project was one of the first applications of NVivo in a non-English context, and probably the first in Japanese. In Japanese, subject and object are often skipped, especially in conversational use. When a phrase of dialogue is coded, and separated from the paragraphs, it is clearer in the English translation than in Japanese what the word or sentence means in the context. NVivo enabled me to locate and compare content not only in the same set of data, but in other types of data, such as TV and newspaper reports, and ethnographic notes and interviews.

The media data was the first to be coded in NVivo; the transcribed documents were coded with tools that NVivo uniquely has. The In-Vivo coding tool applies code to selected words in a text. Not only was this a convenient way to start, but it was also a good way to learn the words used that reflected new concepts born after the earthquake. Even though I am a native speaker of Japanese, especially the dialects of the Osaka-Kobe area, there were new words used that I did not know. If I were to use those words while talking with people from Tokyo or other areas in Japan, they would not understand and might even question if such a word exists. Professor Lyn Richards advises researchers who use In-Vivo to think carefully about the names they use to categorize data; use terms that occur in the data only if they accurately name the category.

The In-Vivo coding tool was a good place to start and helped me to think analytically. However, because this was where analysis began, some nodes that could have been coded under one node at the first coding stage were given different codes at a later stage. I was able to merge these nodes and rename them. While coding, I also made some DocLinks and started to write up memos. I created a new free node with words like 'religion' and 'size' that do not otherwise appear in documents, and hence would not appear in the In-Vivo coding. I coded 'religion' in a PRH field note document and created a new free node 'size' because the size of the building groups seemed to matter. Some concepts were revealed as important when a TV program (NHK) discussed them, such as 'the size of the housing community'. By reading my field notes again on the same day, I noticed that those statements were repeated, such as, 'the large size makes it difficult to develop a good community.'

Other nodes incorporated attitudes. Nodes such as 'positive' and 'negative', and 'plus image' and 'minus image' helped in the analysis of the tone of the media data and how it changed over time. These were incorporated into tree nodes such as TV station, age, gender, housing type, actor, and health issues. Diagram 3.5.2.2 in A3.5.2 illustrates this process.

Japanese verbs reflect a hierarchical society, and vary depending on whether the speaker is superior, equal or inferior to the audience. These help to indicate who is talking about whom without using the subject and object. With hierarchy built into the language, it is possible that a different concept of autonomy is needed. How can ideas like democracy and autonomy work in such a system? This was not the subject of my research, rather it was the context within which I worked and had to make sense of community activity.[5]

Harre (1998) describes the relationship between Japanese language and social relationships in the following passages:

> The Japanese language is tied into and encodes a much more complex system of social relations than almost any other. Indeed the Japanese social order is so fully encoded in grammar that accounting must always take place within an effectively non-negotiable grid of condescension and deference (Harre, 1988: 160).

> In Indo-European social orders it is enough to know who is the speaker and who is the person addressed. But in the Japanese social order we need to know what is the relevant speaker's group and the group of the person or persons addressed (Harre, 1998: 161).

In Japanese, the verb form varies depending on the relationship between the speaker and their audience. It is often possible to understand the context of a dialogue and the relationship between the participants by the verb alone. For example, a verb can indicate if the context of their relationship is hierarchical, formal or informal, and may indicate if the relationship is friendly, neutral or distant.

Fragmentation of data is perhaps the most significant weakness in computer-assisted qualitative data analysis (Hollway and Jefferson, 2000: 68). This is even more a problem in Japanese than in English because of the language characteristics explained above. When a sentence or phrase of dialogue is cut out by coding and separated from its context, the meaning is clearer in the English translation than in Japanese. To minimize fragmentation, I referred often to the data without assistance from the software during the process of analysis.

**Translation**

Although I did my fieldwork in Japanese, I typed up my field notes in English because I wanted to show them to my supervisor and to present them to an English audience. When I translated my field data into English, I tried to preserve the elements of the data that conveyed characteristics of the Japanese language explained above. However, after the field data had already been translated into English, it might have been difficult to identify the degrees of relationship between groups and individuals, such as closeness, seniority, politeness, the degree and direction of respect and where the speaker stood in the relationship. However, the chance of losing something of the essence of social relations in analysis and interpretation was minimized by doing all the English translation of the field data, the coding and analysis of the data myself.

**Coding**

Using color coding, I coded words pointed out by my English supervisor. For example, if she asked, 'What does "loneliness" mean in the Japanese context?' I could use the coding to develop a discourse of loneliness in the Japanese context. Looking at Japan from the Western viewpoint is an important aspect of the research project and different people code data different ways. When I participated in the NVivo seminar, my data was given to the group to code. It was very interesting to see how different each person's coding was. For example, Professor Richards selected words like 'burden' and '*kodokushi*', which means isolated death. She asked me to explain how bad it is in Japanese culture for a person to die alone. This is an interesting example of looking at Japanese values with Western views. She also asked me to explain the comment of the Director of Health of Kobe City, 'Women are doing ok. The problem is men. Imagine what will become of me if my wife dies!'

Another seminar participant, a researcher on housing and care in Australia, felt that community development through origami class was a strange idea. In Japan, this is a common practice.

**Text search**

Once imported into NVivo, I was able to use text search to locate ideas and get a quick overview of the whole document in a short time before coding line by line. Text search was also useful for, for example, accessing the node of a key person's name or title to understand and compare their philosophy, and revealing consistencies or contradictions from different times or different types of data. Some people I met during fieldwork also appeared in TV or newspaper coverage. I interviewed a TV reporter who appeared in a TV program on 31 May 1999. The meeting went for one hour and 45 minutes and addressed what could and could not be said on air, among other things. By making comparisons, things that did not come up clearly on the first occasion would come up again later and so lead to better understanding.

As a part of the analysis, I made a memo named 'Diary to Supervisor' to record the process and note issues to discuss with my supervisor. This was date and time stamped and included ideas about what to do next so that I would not forget them. I kept going back to the data from different angles when writing up reports and chapters and NVivo helped me to do this through the text search feature. For example, for chapters five and six I searched for nodes like 'community development', '*renkei*', 'high-rise', 'independent living', 'support each other', 'lunch', '*yūai* home visit'. For chapter seven I searched for nodes like '*fureai*', '*tsunagari*', 'happiness', 'human contact', 'living alone'. For chapter eight I searched for nodes likes '*kodokushi*'. A2.3 presents these trees and nodes. Writing up was a continuous process of analysis and qualitative references were consulted (Wolcott, 1990 and 2001, Becker and Richards, 1986, Ely et al., 1997, Woods, 1999 and Richardson, 1994).

## Summary and discussion

Secondary analysis of the Hyogo survey data revealed two changes over time: first, the state of the disaster survivors, and second, the points of focus that reflected what officials thought were important. The media coverage also showed changes over time in the topics that were presented. The three sites visited revealed what was happening on the ground and confirmed that one set of myths about TSH was only partly true and that PRH were far from simple solutions to the problem of housing survivors.

Methodological triangulation, a process of combining quantitative and qualitative approaches (Robson, 2002: 174), was used to compare media and field data when analyzing discourses on loneliness for chapter seven. The media data was also related to interviews with journalists and other media people. It was also employed to analyze surveys, media and observations in community research contexts.

Case studies from the media revealed how ideas about loneliness and *kodukushi* were built up from very little into new aspects of culture. Analysis of my fieldwork data enabled me to produce an alternative discourse to the media's on these issues. Like the media, the sense of isolation and loneliness was prevalent in TSH and PRH field sites, but this was partly due to the timing of the research, with TSH closing down after the peak of their activity and the PRH only beginning to generate community. In the PRH there was more of a forward looking discourse but only a longer time frame would have allowed any conclusions about whether or not loneliness would decrease over time.

Although the research was timed to correspond with the end of TSH, it was too late to follow the process of community generation in TSH closely. The analysis of the peak of community at TSH was mainly retrospective; observations of TSH activities were conducted after their peak. This limited to some degree the conclusions that could be made about community in TSH. On the other hand, because of the timing, some reports on TSH from *gyōsei* were available for analysis, while there was little published material on PRH. In conclusion, the timing of the research affected the methods used and the process of analysis.

My access to Haruyama TSH depended on developing a relationship with a leader who would give me a role, even though I was an amateur volunteer. At another major TSH, the leader believed that amateurs could not help without training and it was not easy for me gain access other than as a one time visitor. This major TSH could have been a good fieldwork site; it was frequently covered by the media and would have made for interesting comparisons with media data. However, because Haruyama TSH was located in the same administrative district as the Natsu Aki PRH, I was able to follow individuals as they moved from TSH to PRH. This was also the only area where Department of Health of West Ward of Kobe City produced a report. This selection of this site increased the validity of my findings because I could employ data triangulation using observation, interviews and documents such as public reports (Robson, 2002: 174).

I was not able to obtain income data at my fieldwork sites because it was deemed to be a violation of privacy. Asking such a sensitive question could have endangered the trusting relationship that I had established with the residents and the volunteer public health nurses. The residents of these housing estates were low-income and a larger proportion lived on welfare than on pensions. Chapter one presented the average income data for elderly people in Japan.

One of the questions that future research needs to explore relates to the psychological implications of social security and the pension system, especially for the elderly poor. As shown in chapter one, this was a question that was posited as background to the research, but it was not possible to collect data to answer it. This was because personal finance is a very private and sensitive issue, especially when one of the main entitlements to be a resident in public housing is low-income. Elderly people might prefer to say that they have contributed to society and are therefore are entitled to a pension rather than welfare.

I did not have the methodology to collect the details of the pension or welfare that each individual received. The pension system is very complicated and requires a detailed employment and family history. The people in this research were not part of the more affluent population; they were more often on welfare than on pension. Those with a pension were proud of it, seeing it as an acknowledgement of their contributions to society in their earlier life. They might not have been as happy to announce that they had become part of the elderly cohort that some see as a burden on society, rather than a contributor. Those people who lived on welfare often would be unlikely to announce their situation. This was a taboo area for a research student to go into from the beginning. The welfare commissioners who were rather open about telling stories, sometimes off-the-record and with trust, nevertheless were silent about the residents' status of receiving welfare. A new member of staff employed by Kobe City to advise PRH residents on how to manage their household finances and to collect unpaid rent was worried about how to deal with this sensitive issue.

The research sample was not representative of the elderly demographic in Japan. Rather, it was a sample focused on lower income elderly people. By studying this group, some conclusions can be drawn about the future of elderly people in Japan. Poorer elderly people living alone in urban areas will be an increasingly important group in Japan's ageing society.

# *Chapter Four*
# *Quantitative analysis*

Chapter three presented the research settings including descriptions of the research sites for the study. This chapter presents quantitative analysis of aspects of the earthquake and its aftermath. It has two sections: the first presents the secondary analysis of the Hyogo Health Survey over three years, the second introduces the quantitative aspects of media analysis. The qualitative content analysis will be presented in following chapters.

## Hyogo Prefecture Health Survey

The Hyogo Prefecture post-earthquake Health Survey was carried out by the Health Department of Hyogo Prefecture annually from 1996[1] to 1998. The objective of the survey was to understand the mental and physical health status of those affected by the Hanshin earthquake in 1995, to identify the health-related issues and to strengthen the healthy life support policy. The survey was conducted in October each year and the data was published as a report book in March. The aim of the analysis was to compare the results with data from previous years, particularly in relation to findings about being 'old'. The sampling method was discussed in chapter three.

### *The questionnaires*

Although the surveys consistently collected data on demography and other social indicators, questions concerning some more specific health issues changed over time. In the 1997 and 1998 surveys, questions were modified based on the findings of the 1996 survey, including the answers to open-ended questions, the results of other questions and the overall feedback from public health nurses working in the field.

The main issues identified were problems with mental health, alcohol dependency, nutrition and diet. It also drew attention to the home visit program and other health-seeking behaviors. These issues reflected the fact that daily diet is of high concern for Japanese people; good nutrition is seen as basic to the maintenance of one's health. They also illustrate the concern for the eating habits of elderly people living alone and the potential for this to negatively affect their health status.

New questions on support and care by the public administration were added to the survey in 1998, reflecting the importance that the public administration places on adapting health and welfare policy to the experience of the post-earthquake reconstruction process. These questions also showed concern for the maintenance of social networks and the support of healthy communities in new housing developments, especially in the large PRH apartment buildings the housed many elderly people.

The major weak point of the survey reports is that they did not analyze gender to any great extent. Although they collected data on the sex of respondents, the reports did not analyze sex as a variable and graphs were not drawn up by sex. The conclusions focused on old age, but not on gender.

### *Comparison over the years*

A4.1 shows the comparison of items included in the Heath Survey over the years. It illustrates the shifts in understanding, perception and concern. It also shows how over the years the items have become more focused on issues such as support systems and their impact on health.

Throughout the three years, the surveys measured demographic and social indicators, health awareness, mental health and other general earthquake related issues. While in 1996 and 1997 mental health status was measured with the PTSS-10 and GHQ30[2] scales, in 1997 the measurement was changed to the IES-R[3] test for the degree of depression. In 1996 an open-ended question asked respondents to comment on any issues relating to post-earthquake life, but this was omitted from the 1997 and 1998 surveys. A question about coping with stress was asked in 1996 and 1997 but was omitted in 1998.

A question on post-earthquake support was asked only in the 1996 survey. In 1998, the third year, two questions on support were asked, the first being about home visit support by public administration and volunteers and the second was about social support. These two questions were related to specific responses given to open-ended questions in the first year. However, the meaning of 'support' might have changed in this time; the support in the first year related to the recovery process in the short time after the disaster, whereas the support in the third year related to more long term support systems for elderly people living alone.

Alcohol use, nutrition, diet and eating habits were covered in the 1997 and 1998 surveys. Alcohol dependency was often taken up in media coverage. In Japan, eating habits receive attention in general and the Health Department pays considerable attention to nutrition, diet and eating habit education in their programs. When an old person is living alone, adequate nutrition can be difficult to maintain, especially in a suburban area away from markets

Figure 4-1: Proportion of those living alone at TSH and PRH from 1996 to 1998

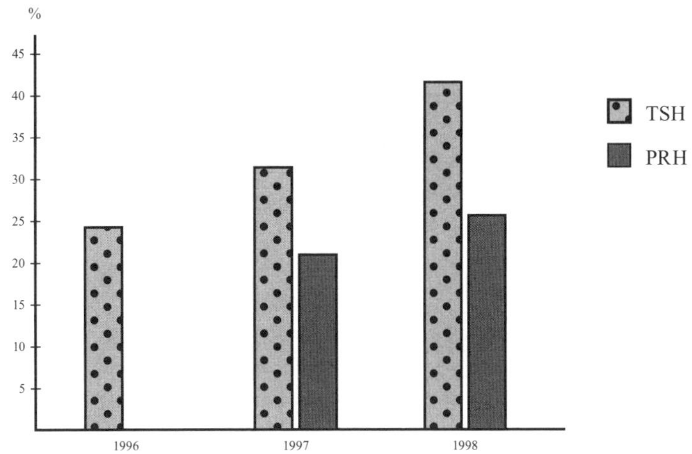


and restaurants. The survey shows how this attention was heightened in response to the earthquake, a point that will be discussed in chapters five and six.

In 1998, new questions on health awareness and habits such as exercise and smoking were added.

**Health trends by housing, 1996–1998**

Figurc 4.1 shows that thc proportion of those living alone increased between 1996 and1998 at both TSH and PRH.

Figure 4.2 shows that the proportion of those reporting sickness remained the same in each type of housing, but the proportion was higher in TSH and PRH than that of those living in general housing. The highest was observed among TSH residents. This could be because the sickest were left behind, because TSH made one sick, or because of something else.

Figure 4.3 shows the reported sickness of elderly people by age group and housing type in 1996 and 1997. People aged between 70 and 80 reported the highest proportion of sickness. Those aged 80 and above were healthier than those aged between 70 and 80 with the exception of those that lived in TSH. In TSH, the eldest group was sicker, particularly in 1996. The 1997 data divided the 60–70 age group into 60–64 and 65–69. The age group 60–64 was proportionally healthier than the 65–69, which contained the highest proportion of sick people.

Figure 4-2: Reported sickness by housing type from 1996 to 1998

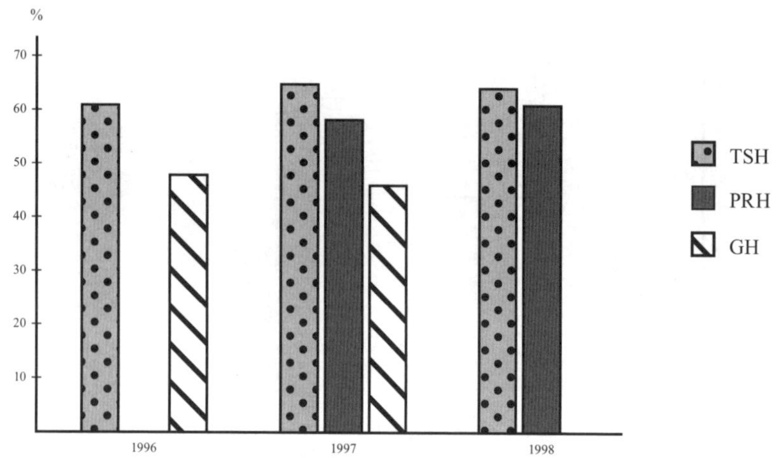


Figure 4-3: Reported sickness of older people by age group and housing type in 1996 and 1997

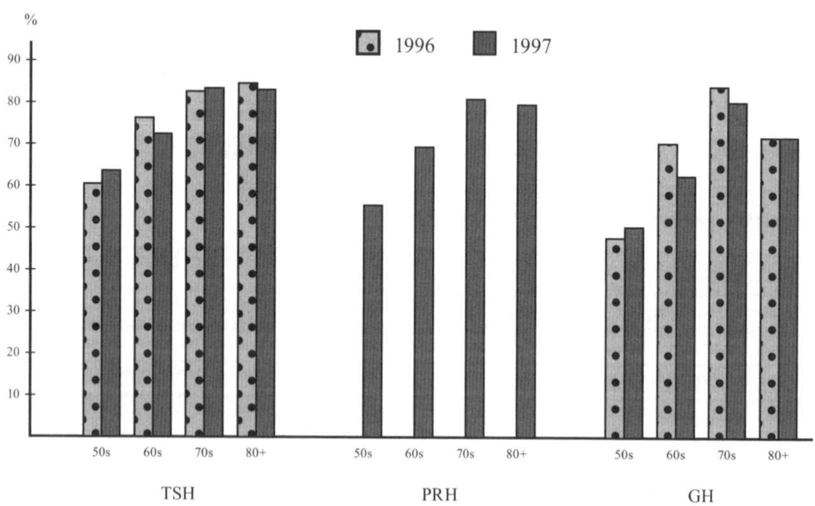


The Mental Health GHQ 30 test is used to measure general mental health. The average score for Japanese people is 15–20%. The sample population reported about twice the Japanese average, scoring 35.6% in 1996, and 36.8% in 1997 (Figure 4.4). In all types of housing, people aged between 50 and 60 reported the highest proportion of problems with mental health.

Figure 4-4: Mental Health GHQ30 Test by age group in 1996 and 1997

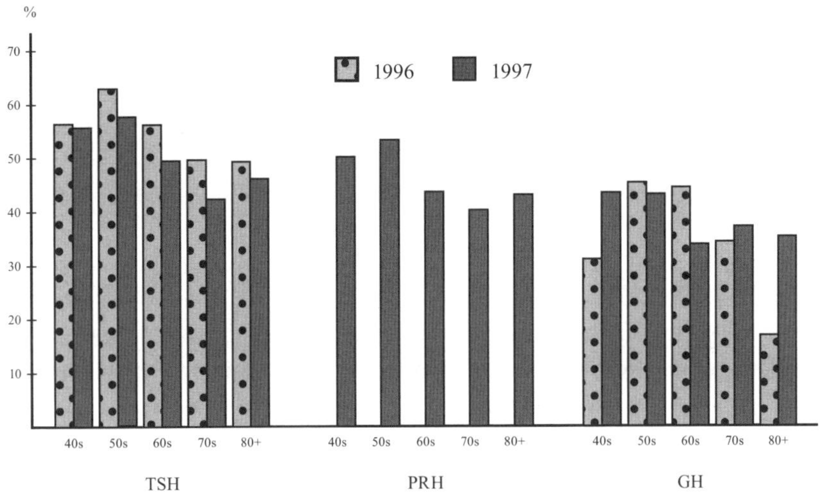


As the test was changed to the Diagnostic and Statistical Manual of Mental Disorders 4th edition (DSM-IV) in 1998, the trend over the three years cannot be simply compared. Different tests could result in different findings. For example, the study on adults aged 65 and older who survived the 1989 earthquake in Newcastle, Australia, showed completely different findings on various tests including the GHQ and IES (Ticehurst et al., 1996).[4] In the 1998 Hyogo Survey, TSH residents aged between 40 and 50, and 60 and 70 were reported as having the highest rates of depression (12.6–13.0%), while in PRH it was people aged between 50 and 60, followed by those aged between 40 and 50 and those aged 80 and over (7.7%) who scored the highest for depression.

When the GHQ 30 test is applied to other cultures, the result might not be reliable. Figure 4.4 shows that mental health improves in better housing. This finding is same as the study from the earthquake in Umbria, Italy in 1997, although a different test was used. People living in prefabricated huts showed a higher score on both the Geriatric Depression Scale and the Hamilton Anxiety Scale and complained more often about their health status when compared with home dwellers. While all of the participants suffered from the discomforts caused by the earthquake, the precariousness of living in temporary houses could explain the higher distress experienced by those housed in huts (Mecocci et al., 2000).

**Alcohol dependency**

The Kurihama Alcoholism Screening Test was developed to measure an individual's drinking behavior in the Japanese societal and cultural context (Morioka et al., 1996). It is a set of 14 questions like 'Has a person experienced a problem with a significant relationship because of their drinking habit?' and 'Even when determined not to drink, are there times when a person cannot help drinking?' In Japan, the average proportion of people with serious problems with alcohol is 5.0–7.1% for men and 0.3–0.6% for women.

In the Hyogo survey of 1997, the proportion of residents having serious problems with alcohol dependency in was 7.0% in TSH, 5.5% in PRH and 4.7% in general fousing (Figure 4.5). Alcohol dependency was highest among TSH residents and lowest among general housing residents. It was higher among men than among women. Also, the problem is more serious among men living in TSH than men in PRH. It was least serious among the men living in general housing. There was no significant difference among women across different types of housing. Middle-aged residents, aged between 40 and 59, had more serious problems than older groups. Looking only at men, the degree of the problem was more serious among TSH residents.

As Figure 4.5 shows, alcohol dependency increased moderately in 1998 from the previous year in both TSH and PRH. The highest scoring group in TSH was aged between 50 and 59 (21.5%), while in PRH the highest was aged between 40 and 49 (15.8%).

The surveys indicated that the health status of the residents in TSH was overall worse than the health status of the residents in PRH. This could be explained by living conditions being more difficult than in PRH, or perhaps people with poorer health tended to prolong their stay in TSH. It might be a combination of both explanations.

Residents aged over 65 received particular attention in the survey. When interviewed about elderly residents of TSH and PRH, a high-level of public health officials at the local government level said, 'Women are doing ok, the problem is with men.' However, as the findings were presented, there was no analysis of the gender variable.

## Media data

As explained in chapter two, media analysis was a main part of the research project. Chapter three introduced the methods whereby media data was collected. This section presents the findings from the quantitative analysis of the media data.

Figure 4-5: Alcohol Dependency KAST Test by gender and housing type in 1997

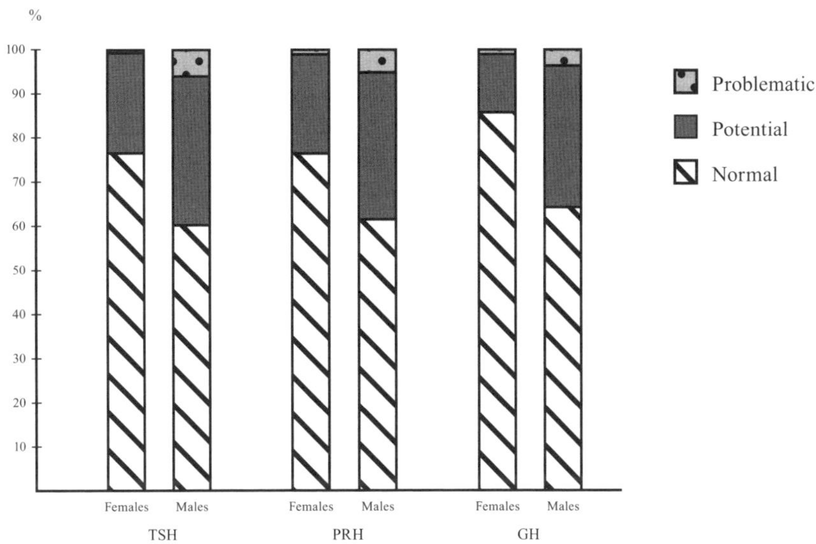


### *Time allocated to elderly people*

The broadcasting time allocated to elderly people decreased from 9 hours and 10 minutes (33% of the total broadcasting time covering earthquake related topics) in January 1999 to 3 hours and 20 minutes (8.4%) in January 2000. Table 4.1 shows the length of time given to broadcasting issues related to elderly people in proportion to the time allotted to the earthquake in general. Elderly people were seen as an important population in the context of the ageing society, but attention shifted to other populations and issues as well.

### *Gender*

In television documentaries and news programs, more elderly women than elderly men were featured (Table 4.2). Programs generally featured one man and one or two women. That may reflect the fact that more elderly women than elderly men lived in these communities, as the media probably tried to cover both men and women.

### *Positive and negative images of housing*

While they depicted unhappy and lonely elderly women as well as elderly women with positive attitudes, all the TV stations, public and private, national and local, presented elderly women's experience as positive in both

Table 4-1: Television coverage of earthquake related topics in 1999 and 2000

| | 1999 Jan | 2000 Jan |
|---|---|---|
| Total time | 27 hours 55 min (1,675 min) | 38 hours 55 min (2,335 min) |
| Time covering elderly people | 9 hours 10 min (550 min) | 3 hours 20 min (200 min) |
| Proportion of time covering elderly people | 33% | 8.4% |

Table 4-2: Television coverage of victim's lives by gender and station in 1999 and 2000

| | 1999 January | | 2000 January | |
|---|---|---|---|---|
| | Male | Female | Male | Female |
| NHK Ch.2 | 6 | 12 | 0 | 0 |
| NHK Education Ch.12 | 0 | 0 | 3 | 5 |
| NHK BS | 0 | 0 | 0 | 0 |
| Mainichi MBS Ch.4 | 6 | 12 | 1 | 1 |
| Asahi ABC Ch.6 | 3 | 2 | 0 | 0 |
| Kansai Ch.8 | 0 | 2 | 2 | 2 |
| Yomiuri Ch.10 | 5 | 10 | 2 | 1 |
| TV-Osaka. Ch.19 | 0 | 0 | 0 | 2 |
| Sun-TV. Ch.36 | 3 | 5 | 0 | 1 |
| Total | 23 | 43 | 8 | 12 |

Note: Three men were chief of the self-governing body in their community: one man on NHK and one man on Sun-TV in 1999, and one man in Mainichi MBS in 2000.

1999 and 2000. Usually, an interviewed woman was presented with company, often with a group of women friends. On the other hand, men in TSH or PRH were always presented as unhappy unless they had the company of a wife.

After a January 2000 Sun-TV news program on life in PRH communities, a male newscaster commented, 'Men, let's work harder and more positively like women are doing.' Elderly women made friends and looked happier than men who appeared inactive. A male resident in PHR said in an interview,

Table 4-3: Television program perception of housing types in 1999 and 2000

| | Positive | | Negative | |
|---|---|---|---|---|
| | 1999 | 2000 | 1999 | 2000 |
| TSH | 2 | 1 | 3 | 3 |
| PRH | 3 | 3 | 12 | 5 |
| Total | 5 | 4 | 15 | 8 |

'Women are doing fine; they just keep talking and make friends. But for a man, when he is old, it is difficult to make friends just sitting in front of the TV all day.' See chapter six on community involvement and gender for further discussion of this issue.

Although the tone and focus of PRH TV coverage was negative in 1999, in 2000 it shifted to more neutral and even encouraged people to think positively about life in PRH (see Table 4.3).

In January 2000, in the fifth year after the earthquake, the whole media became more positive towards the future for PRH rand TSH residents. The media tried to take a brighter perspective on all housing types, while still trying to report the problematic issues. In January 1999, it was more a case of finding and reporting negative issues with an overall negative tone (see Table 4.3).

## *Kodokushi* (isolated death)

*Kokokushi*, meaning isolated death, was used by the media as a sensational and attention-grabbing word to describe a post-earthquake problem. According to *Asashi Shinbun* newspaper report on 24 April 1998, 207 isolated deaths had been reported in TSH, excluding suicides.

Doshisha University Life Issue Study Group studied the background of the isolated death cases occurring in TSH (Doshisha University, 1997). The causes of isolated death varied. In addition to common causes of death among Japanese such as heart disease, other rarer causes like starvation and malnutrition were reported. These are not only related to poverty but also to alcohol dependency and mental health problems such as hopelessness for their future. The number of isolated deaths was 72 for the first year and 70 for the second year. As more and more people were moving out from TSH communities and those remaining had fewer neighbors for company, the rate was rising.

The number of isolated deaths of males was double that of females. The average age of the isolated deaths of males was 55, for females it was 70

Figure 4-6: Distribution of isolated deaths by age and sex

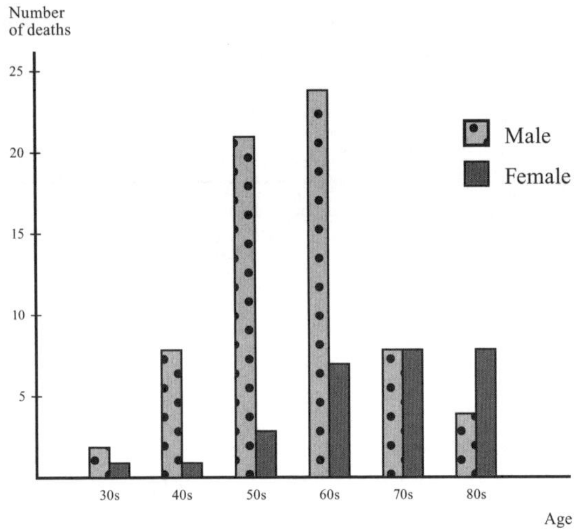


Source: Ueno (1997: 146).

(Figure 4.6), while life expectancy of Japanese males is 77 and of females it is 83. Isolated deaths of males in their 50s and 60s accounted for nearly half of the total isolated deaths. This is the same gender and age group that was the most vulnerable in the former Soviet Union after the political and economic transitions of the early 1990s (Goldstein et al., 1996: 9). There is no baseline figure to compare rates before and after the earthquake because TSH schemes were built after the earthquake.

Immediate causes of *kodokushi* included alcohol-triggered liver diseases (43.8% of male cases aged between 40 and 60) and malnutrition (Ueno, 1997: 150). The average age of k*odokushi* for women was much higher than that of men. The main cause of *kodokushi* among those aged over 65 was heart disease. More cases of *kodokushi* from heart diseases were reported among women (Ueno, 1997: 151).

Qualitative aspects of *kodokushi* will be discussed in chapter eight.

## Summary and conclusion

Chapter two showed that there has been little study of the long-term impacts of natural disasters world-wide, with no ethnographic research published. It also showed that elderly people were the main victims of the 1995 Great

Hanshin Earthquake in both mortality and morbidity; they received more attention from academics in media and journals than other age groups.

This chapter presented secondary analysis of data from the Hyogo Prefecture post-earthquake Health Survey. This data on key health indicators is background information about the study population. Although elderly people are less healthy in general, the Hyogo Health Survey results show that the population in the 50s age group, or even in the 40s for some cases, reported more health problems than their older cohorts. The second section of this chapter showed that the earthquake related media focus, and subsequently the public attention, was on elderly people.

The analysis of questionnaire change over the years showed the shifting focus and values of the public administration. This is an important point as questionnaire responses are often taken as a basis for policy development. As presented in chapter one, Campbell (1992) showed that the Japanese media has played a crucial role in health policy change for elderly people. The survey results, as is often the case, were not analyzed well. Rather, respondent's answers were simply incorporated in the following year's survey. The questionnaire changes reflect what the Department of Health were learning from the survey results of previous years, or from their experiences working within the localities.

This chapter has shown that from the early stage in the aftermath of the 1995 Great Hanshin Earthquake, the attention of the *gyōsei* (public administration), media, and researchers was on elderly people affected by the disaster. The data on the short-term health impacts of the earthquake on victims have shown that a larger proportion of victims were elderly people. These figures support the need to focus on elderly people. The attention given to the elderly population is consistent with the attention given in Japanese society to population ageing. The measure of earthquake effects shifted from short-term physical damage to mental health, but the measures were crude and not all related to Japanese culture. To cope with the ageing society, different measures are needed.

The second section of chapter four looked at the focus and tone of media coverage. The following chapters will continue with the qualitative analysis of media data and the findings of ethnographic fieldwork.

# Chapter Five
# *Temporary Shelter Housing*

## Introduction

As shown in chapter four, the proportion of TSH (temporary shelter housing) residents who were elderly had increased over time. On 17 January 1998, three years after the earthquake, *Asashi Shinbun* newspaper reported that of the approximately 25,000 people still living in TSH, about half were elderly people living alone. The proportion of people aged over 65 in temporary shelters in Hyogo Prefecture was 30.3% in February 1996, one year after the earthquake. By 1998, the average age of TSH residents was over 65 years old. There were more women remaining in TSH than men. According to a report by *Asashi Shinbun* newspaper on 2 April 1998, these demographic features of TSH communities were similar to the projections for the whole of Japan in 40 to 50 years.

## Background to TSH fieldwork

My fieldwork took place in the Natsu Aki PRH, the Fuyuyama PRH, and the Haruyama TSH (which is near the Natsu Aki PRH), although I visited three other PRH and two other TSH sitess which I also referred to in my analysis.

The Natsu Aki PRH and the Haruyama TSH were in the West Ward of Kobe City. The Fuyuyama PRH was in the Central Ward of Kobe City. Although the West Ward was not directly hit by the earthquake, it was the area most affected by the post-earthquake reconstruction. The population of the area increased with the relocation of people to the temporary shelters, most of which were in the West Ward (see Table 5.1).

With three compounds, Haruyama was one of the largest TSH schemes (see Table 5.2). The first compound had 66 households, the second had 750 households and the third had 500 households. In May 1996, the proportion of those aged over 65 was 44.2%, 26.3%, and 7.1%, respectively. The higher proportion of elderly people in the first compound might reflect the fact that the priority of allocation was given to elderly people. The buildings were single storey apartment style housing (see Photos 5.1 and 5.3 for photos of the TSH building style).

Table 5-1: Population of the West Ward of Kobe City before and after the earthquake

| | Total 1 October 1994 | In TSH 24 May 1996 | Total 1 February 1996 |
|---|---|---|---|
| Population | 199,951 | 11,116 | 213,531 |
| Households | 63,051 | 4,437 | - |
| Aged 65 and over | 15,796 | 4,437 | 20,428 |
| % aged 65 and over | 7.9 | 39.9 | 9.6 |

Source: Hyogo Prefecture College of Nursing (1996: 7).

Table 5-2: Population of Haruyama TSH in May, 1996

| Haruyama TSH | Number of households | Total population | Population aged 65 and over | Proportion 65 and over |
|---|---|---|---|---|
| I | 66 | 95 | 42 | 44.2 |
| II | 750 | 1,108 | 291 | 26.3 |
| III | 500 | 785 | 56 | 7.1 |
| West Ward total | 8,941 | 11,116 | 4,437 | 39.9 |

Source: Hyogo Prefecture College of Nursing (1996: 10).

This chapter presents fieldwork from the Haruyama TSH site, which I conducted in the second compound. Background on PRH sites will be introduced in chapter six. Figure 5.1 illustrates two typical designs of TSH apartments.

## Analysis

The TSH scheme employed *nagaya* type housing consisting of one floor hutted apartments. One building had five to ten households with one or two rooms each.

The majority of those left behind in TSH were elderly people. Those who had moved out from the shelters often found their new living situation to be unsatisfactory, and repeated relocation is not easy for elderly people. New homes were often located in inconvenient suburbs far away from where the residents used to live, making it difficult to commute to medical facilities

Photo 5-1: Suburban type TSH complex surrounded by mountains

Photo 5-2: Bridge over the beautiful green valley on the road from the subway station to Haruyama TSH. Some TSH residents jumped from this bridge to commit suicide.

with which they were familiar. Elderly people might have felt particularly isolated because accessing social services and activities took greater effort for them. Just meeting basic needs such as grocery shopping was hard for elderly people when they often did not drive.

Some who did not move into TSH moved in with their relatives, often causing stress to both sides when their stay was prolonged. Some elderly

Photo 5-3: ‘No entry’ sign put up on the fence of the TSH

Figure 5-1: TSH design Type 1K (1 room (6 tatami-mat size), with kitchen, and bath and toilet room) and Type 2K (2 rooms (6 tatami-mat size and 4.5 tatami-mat size) with kitchen, and bath and toilet room).

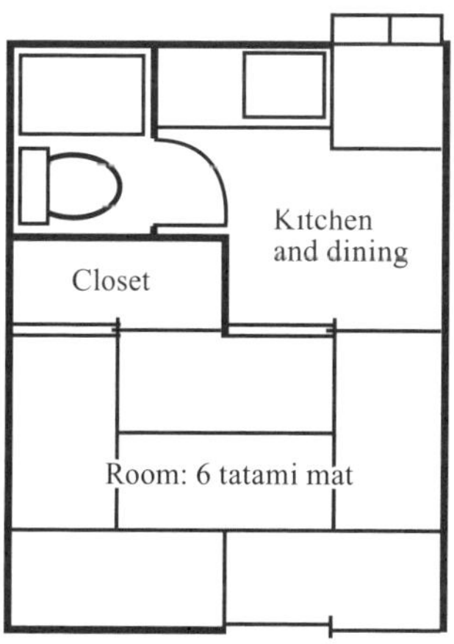


1K type: kitchen and 1 room

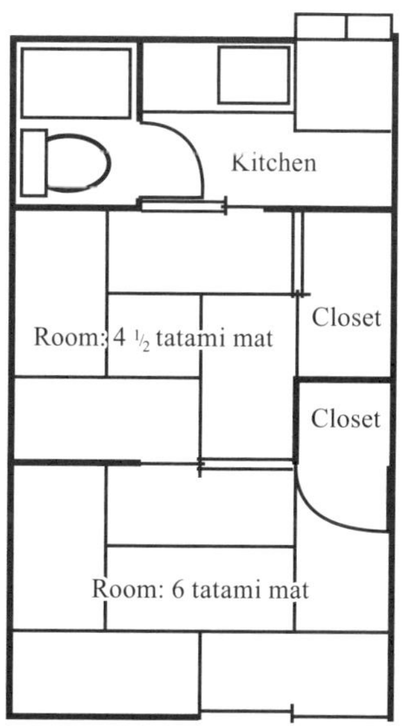


2K type: kitchen and 2 rooms

people who were hospitalized kept finding medical reasons to stay in hospital as they had no home to return to. Some neither moved into TSH nor were hospitalized. It was difficult for volunteer workers to reach those who had moved to a new place other than TSH because they were spread out among people who were generally more affluent.

In accordance with Japanese values of fairness, TSH units were allocated by lottery. In the early rounds, the lottery had a quota for elderly people. The giving of priority to elderly people reflected the values of Japanese society, but it can also be understood in the context of the tendency to group the elderly with *jakusha* (weak) and disabled people. By concentrating in one area of TSH the elderly and weak, the administration created an isolated area of people with greater needs (Ogino, 1999: 340–341).

Elderly people were reported as having a more difficult time adjusting to life in the evacuation centre (see chapter one). Often, they did not want to risk waking their neighbors by using the temporary toilet outside in the middle of the night, and they often did not eat well or drink enough water. Japanese elderly people in general are often not outspoken and try not to be a burden to other people. They do not generally draw attention to their troubles and seek help unless people come and ask privately and on an individual basis. Elderly people were not the first to line up for water, food and other emergency supplies when these were delivered to the evacuation centre. For these reasons, giving priority to elderly people in the housing lottery was considered fair in Japanese society.

The same lottery system was used to allocate PRH apartments. Those that won an apartment in an area they wanted were happy, whereas those that did not were unhappy. Many applied for the second choice and third choice in the next rounds and some committed suicide when they received the refusal letter in the third lottery. Others were unhappy but took their loss as bad luck.

### *Characteristics of TSH residents*

Typical characteristics of elderly TSH residents include living alone, receiving little if any pension, owning no home and alcohol dependency, especially among those who were late in moving out. Many residents shared more than one of these features.

Despite sharing many of these common features, elderly people living alone in the shelters had different backgrounds. Some had never married. Some lost their spouse in the earthquake. Some got divorced because of the earthquake or after hard times caused by the earthquake. Some were widowed before the earthquake. Some had no children or their children could not afford to have them in their homes due to the limited space or the possibility of conflict. Some children of elderly TSH residents visited their parents frequently and helped them, while some others never visited.

The tendency of administrators to treat elderly people as one homogeneous group was not often criticized. The elderly people in my study had diverse backgrounds, despite sharing some characteristics like nationality, region, language, education, employment and class position. During my fieldwork, I tried to find out the reasons why some elderly people who had children did not live with them. For many, family relationships can be sensitive and many did not want to talk about it. However, I learned from people who worked in the community that some children could not afford to house their impoverished parents because they themselves were poor. Other parents were reluctant to be a burden on their more affluent children. Because of the high-levels of social mobility and educational opportunities in post-war Japan, many elderly parents feel they belong to a different class to their children.

TSH residents lived in close contact with their neighbors and sharing food was a way to meet people and make friends. Such food exchanges were also important for maintaining nutrition and a balanced diet, especially for those who lived and ate alone. Eating habits will be further discussed in the following chapter as they are important in Japanese culture and values.

Some TSH residents were good at organizing themselves and negotiated with the public administration to enable them to move together to the same PRH compound. This was not for physical health reasons, but to enable them to stay close to their friends and neighbors in their new homes. In an interview that I conducted with the Director of the Department of Health and Welfare of Kobe City in 1999, they said that compared to the problems associated with allocating people to housing by lottery, this was as a success story.

### *Problems*

Elderly residents of TSH lived in isolated hutments. Because of the well intended scheme of giving elderly and disabled people priority in the distribution of temporary housing, a community of elderly and disabled people living alone was established (Tanida, 1996: 1133-5). Furthermore, because they were given priority and assigned early on in the process, their houses were located at the end of the TSH area, farthest from the community center and the main entrance (Figure 5.2 shows a map of the Haruyama complex and bus stop). Sometimes it took one hour for an elderly person to walk from her shelter to the main entrance to catch a bus from the street, although the same distance might take only ten minutes for a younger person to walk. Also, the plan to giving priority to elderly and disabled people concentrated those who needed

Figure 5-2: Map of Haruyama TSH complex and bus stop

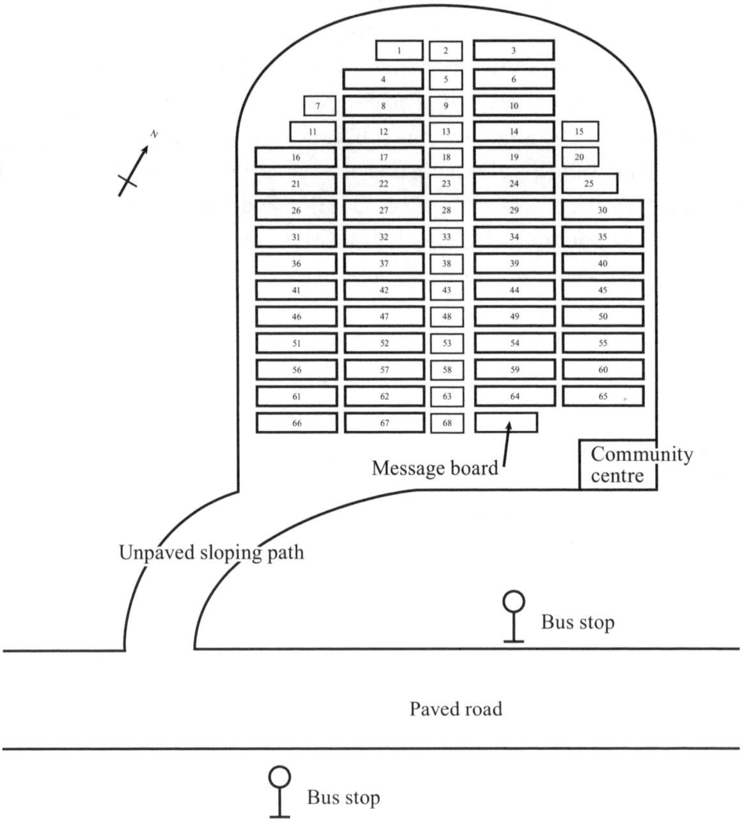


support in one area rather than spreading them out in the community. A survey conducted in late September 1995 showed that 40% of elderly TSH residents lived alone, away from family, in a household designed for a single person.

As those who could move on in their lives gradually left TSH, the shelters were left to the most vulnerable residents, such as elderly, disabled people and low-income families. In June 1996, 49% of the shelters were occupied by families with members aged over 65 years old, and 28% were occupied by a single person aged over 65 (Tanida, 1996). In March 1998, *Asashi* newspaper presented the findings of a TSH health survey conducted by Kobe City public health nurses in autumn of 1997. The report found that the major problems affecting residents were aged dementia, mental and physical disability, elderly people who were *netakiri* (bedridden), mentally ill and alcohol dependent (see Table 5.3).

Table 5-3 Priorities in allocation lottery for Kobe City TSH

*First priority*

Household with only people aged 60 and over

Household with an adult or child classified as 1st and 2nd class mentally disabled (moved from second priority after the first round of invitations)

Single mother household with child under 18 years of age

*Second priority*

Household including people aged 65 and over

Household including infant aged 3 and under

Household with pregnant woman

Household with three or more children aged under 18

*Third priority*

The weak

A household with a person injured by the earthquake

A household with a person physically weakened by the evacuation

*Fourth priority*

Other household not covered by the above three categories

Source: Kobe City Post-Earthquake Reconstruction Office.

Table 5-4: Health survey at temporary shelters in Kobe

| | **Number of people** |
|---|---|
| Aged dementia | 120 |
| Mentally and physically handicapped | 304 |
| The bedridden elderly people | 146 |
| Mentally ill | 85 |
| Alcohol Dependent Syndrome | 144 |

Source: *Asashi* newspaper (1998).

Media comparisons between TSH and PRH have produced conflicting results. Some reports said that life in TSH could cause and worsen health problems (Table 5.4 illustrates the extent of health problems in TSH). Life in TSH was too hot in summer and too cold in winter. The TSH compounds were far away from people's doctors so they could not see them regularly. This increased the daily anxiety of elderly people as health is often a major concern in old age. Many wanted to make sure they had reasonable access to medical services if something should happen to them. However, other media reports did not think that life in TSH caused health problems.

Mr Minamimori, a TSH resident, said in an interview about his group house plan on NHK, 11 January 2000, ‘When I am fit and healthy it is OK to live alone. But if something happens...you know...’. In Japanese society, the health clinic is often seen as a main venue for social occasions for elderly people. After elderly people receive the free medical services at the health clinic, they go to the waiting room to meet with other elderly people. A famous joke in Japan is set in the waiting room of a clinic. If one regular patient does not come, the others ask each other, ‘What happened to them? Did they get sick?’

On 1 January 1999, Yomiuri TV reported that TSH life was adversely affecting (*mushibamu*) the health of Mr Hiroshige Takehara, aged 71. Another NHK TV program, broadcast 14 January 1999, reported that Mr Yasuo Uematsu, aged 70, lived in TSH for two years. A draught in his accommodation worsened his asthma.

As these examples illustrate, TV in 1999 reported that life in TSH was worsening people’s health. But when I did my fieldwork people put an emphasis on the positive effects of TSH life on their health, assuming that I had already heard about the media reports. Some explained that what was reported in the media was true in some parts, but in fact the effects were mostly positive. They said that their health had improved because they lived in a mountainous area with clean air and had to walk long distances. Walking is good exercise to maintain one’s health, they said.

Ms Ueda, a TSH resident, said in a TV program in 1999 that she did not like living in TSH and that she felt lonely to see others leaving. But in 2000, on another program, she looked back on those days in TSH gratefully, saying that she learnt from the experience that people need one another.

### *Elderly people’s happiness in TSH*

The happiness of TSH residents was a common topic of TV programs. In 1999, most programs showed TSH life as happy for elderly people.

On 13 January 1999, ABC TV presented an old woman who had moved to PRH. She said,

> It is six months since I moved to this public reconstruction housing. But I have made no friends here. I do not feel positive any more. I feel a lot of anxiety. A day goes without any chance to talk with anyone. I watch TV all day and just sleep.

In a report on ABC TV, 15 January 1999, Mr Kawabata said he had been happier in TSH. Since moving to PRH, he felt lonely. He had no one to talk to and no one to share food with. He found it difficult to make new friends with his neighbors. He said that when you are old it is difficult to make new friends. He saw old age as a factor in his difficulties:

> I have no energy left to say hello to my new neighbor. The life in the temporary shelter was much better. Everyone here closes themselves in their home. It is impossible to have any communication. In this kind of place, people just briefly see each other.

This program stated that life in TSH was much better. People came to miss good things about TSH life, but they could only say so after they had moved to a permanent place. After all, most residents wish to move out from TSH. The media was ignoring this fact. Their interpretation was not correct in saying that people want to move back to TSH because PRH was a horrible place.

In another ABC TV program, broadcast 17 January 1999, 76 year old Ms Sachi Yamamoto said about her life in PRH,

> I have no one to make friends with here. I say hello to a neighbor. I go out to participate in a cleaning activity of the housing community. Three months after I moved in, I started to feel not good. I felt depression. It is difficult to express this feeling. When I moved in, there were some deaths, one after the other in this building. Someone jumped from the top of the building to commit suicide, a 67 year old woman living alone. The life at the temporary shelter was good. I miss those days. I often make a visit to my former temporary shelter. I wish they repair the shelter a bit and let me live there.

However, TSH was no longer the same place. Mr Jiro Sezoko, aged 57, said in the same program that although he could not make a comparison as he had not yet moved out, he was not happy living at the temporary shelter. He said he did not feel at ease and that he wanted to move out of TSH while still alive.

Ms Yoko Fujikawa, aged 50, appeared in a few different TV programs. In an interview on 17 January 1999, she talked happily about living in TSH and how her friends who moved to PRH would often return to visit her and say, 'It's better at a temporary shelter'. This may be true for some reasons, but those friends might be showing kindness by saying this so as not to make her envious of those who had already moved out. Did the information Ms Fujikawa received discourage her from moving out? Did it lower her expectation of her future life at PRH? Was she prepared for the new living place? Having friends that continued to visit her was a good support network and kept her happier. Physically able friends could visit each other, but those who were physically weak could not travel long distances to visit one another.

In a NHK program on 17 January 1999, Takeshi, the son of 82 year old Mr Kiyoshi Nagata, described how he had decided that TSH would be best for his bedridden father because he thought there would be people around. He expected that at TSH no one would be left alone and he valued this aspect. He also could not find another affordable place. TSH was rent free, but the media did not report this as if it were a taboo. However, one could live in TSH for only so long as it was opened and this was not for long. In April 2000, the last TSH was closed down.

The same NHK program featured a couple, Mr Kunio Fujihara and Mrs Fumiko Fujihara, who were visiting the TSH where they had previously lived. They said that although the PRH was well-built life there was less happy. They were not living alone but still they missed the human contact (*kōryu, fureai*) they once had with their TSH neighbors.

A Mainichi TV program, 17 January 1999, said that life in PRH was worse than life in TSH. An old woman, who had been living in TSH for some years, felt sad to leave. Two men who had not been able to move out said that they had the impression that those who moved to high-rise PRH were suffering from depression. They had heard of two cases of suicides in PRH, an old woman and a middle age man, and another case of an old woman who went missing when she received the news of winning a place in PRH but did not want to leave TSH.

In media reports in 1999, those who had not yet left TSH talked like they already knew that they would be less happy at PRH. They might have learned this from their friends who had already moved, or from these TV programs. Friends would say things to comfort those still in TSH. They would not say that they were happier in PRH and show off what they had when their friends had not been successful in finding a permanent house or winning the lottery for allocation to PRH. It is also partly because the people still in TSH would say this to ease their own feelings, to tell themselves that their inability to move out from TSH was not that bad.

Another factor in the negative perception of PRH was that TV reported what it wanted to report; it did not necessarily reflect the true feelings of the people. For example, Ms Chieko Uemoto, aged74, said in an interview with Sun TV on 8 January 1998 that she was happy with the PRH that she won after the seventh lottery. This was inconsistent with the general tone of media coverage as detailed above. Because the media perception was generally negative, it is difficult to say how common more optimistic views were among residents.

### *Were elderly people taken care of in TSH?*

Because of the work of volunteers and the public administration, TV reported that elderly people were taken care of better at TSH, despite the poor housing conditions. Elderly residents of TSH may have felt taken care of because of the daily financial and practical help they received, rather than the quality of the accommodation itself.

The media was generally denigrating of PRH in comparison with TSH as housing for an ageing society. However, even though the type of housing was seen as the most important factor in determining the experience of residents, living in TSH and PRH was mediated by programs and community arrangements, not just the housing scheme.

### *Making friends*

The *sawa-kai* (tea party) was a major community event in TSH and this is how it was represented by the media. Such an event was visible and easy to identify as an activity. For a reporter, it was easier than an individual's private apartment to access. Also, it was a good example of a simple event that was very important to the community. It was one of the main research sites and events during my fieldwork both at TSH and PRH. Tea shops were a place to exchange information, to make friends and to go out.

While people may not make a friend with just one visit, I observed people having an opportunity to exchange and start talking after attending a number of times. Although some people who talked one time did not necessarily talk the next, familiarity and friendships sometimes developed over time.

Several residents at TSH seemed to look forward to the weekly tea party as an event they could not miss. Residents at TSH often talked about the weekly tea party when they exchanged greetings. But not all residents found that tea parties helped them to foster new relationships. According to an ABC TV program broadcast on 13 January 1999, Mr Kawabata was happier in TSH. He felt lonely in PRH, finding it difficult to make new friends and blaming his old age for this difficulty.

There were also some reports on attempts to establish collective houses at TSH, efforts that were welcomed by the media. On the 13 July 1999, *Asashi Shinbun* presented arguments encouraging collective houses. On the same day, Mainichi TV ran the headline, 'Group home, a housing scheme good for aged people, suggests experiences at temporary shelters'. In these collective houses, elderly residents did not need the help of younger people, but could take care of each other as each person had a different function, and each could complement and help the other. In this way, each person felt that they were useful to others, not just a burden, thus providing a feeling of *ikigai* (life enrichment, explored further in chapter seven). Also, residents of collective housing generally did not feel pressured to keep up with the pace of younger people.

The other advantage of collective housing is that a smaller number of health care workers and helpers could cover a larger number of elderly households. In this way, the cost of the care of ageing populations could be reduced.

## Analysis of published reports

Tanida (1996) has drawn attention to the elderly people who he says were left behind in temporary accommodation, and argues that special and continuous care is necessary for elderly and vulnerable people after natural disasters. While acknowledging that the Japanese government set up special temporary houses for disabled and elderly people with 24 hour support in late February of 1996, Tanida found that the numbers were too small and many elderly people had to live isolated in ordinary temporary houses.

He also pointed out some problems which he says are unlikely to be solved in the near future. The first problem is the difficulty of community generation among elderly residents of temporary accommodation. Only 30% of temporary shelter areas were able to establish a self-governing body by July 1995. Although this was only a few months after the temporary shelters were built and it was expected to take longer to generate community among residents, Tanida argued that the difficulty in forming community was to an extent produced by the old age of residents.

The second problem was that many elderly people died alone in temporary houses and there was a particularly high rate of suicide. *Kodokushi* (dying alone) emerged as a problem in post-earthquake society and the media used the number of *kodokushi* cases as an indicator of this, a trend discussed further in chapter eight.

Although Tanida's article was written only one year after the earthquake, it identified trends in migration among residents of the temporary houses. Those who were able to leave did so, and over time only the most vulnerable members of society remained, such as elderly and disabled people and low-income families. He also described how those who cared for elderly people before the earthquake were also themselves victims and many were unable to continue care after the earthquake.

Hirayama (2000) argued that the patterns of housing damage in the Hanshin earthquake reflected the geographically divisions which had been growing prior to the earthquake. He argued that while the earthquake itself was a natural phenomenon, the inequality of housing damage was generated socially. In my study, the people left at the temporary shelters and public reconstruction housing were from the lower-income class. Hirayama also pointed out that the process of recovery was not uniform and had deepened socio-spatial polarization and divided the city. He described how temporary housing promoted the grouping of victims and created a group in need of welfare, a high proportion of which were those with low incomes and the elderly.

Hirayami pointed out the problems of temporary housing design in each unit and in the whole area. Differences in floor levels caused problems for elderly and disabled residents. The shortage of common space, green areas and living facilities was striking. I will argue later that the shortage of common areas was also a factor in the difficulties in generating a community.

Hirayami also considered the mental health effects on the residents. He said that many camps were constructed in fenced parks and in sites isolated from neighborhoods by main roads. The word 'camp' is significant here. From my fieldwork, I found that the fenced camp triggered traumatic memories of wartime for many elderly people. Hirayami argued that these circumstances caused residents to become depressed, citing as evidence the increase of alcoholism, lone deaths, suicides and accidental deaths among middle-aged men and malnutrition among elderly women.

Finally, Hirayami explained the distinction between ministries responsible for temporary housing and public housing. Rent-free temporary housing was provided by the Ministry of Health and Welfare, based on the same idea as public assistance. Public housing was provided by the Ministry of Construction.

In the following two special reports on TSH and PRH communities prepared by the Department of Health and Welfare in the West Ward of Kobe City, no distinctions were made on the basis of gender but rather on age and single occupancy. Elderly people were categorized as 'old'. 'Aged over 65 and living alone' was a category describing the most vulnerable population, such as the disabled who required special attention from the public support system.

The Hyogo Prefecture Nursing College (1996) report presents research conducted from April 1995 to March 1996 analyzing support activities by public health nurses. It begins by saying that *kodokushi* was seen as a problem by the media and that the *gyōsei* (public administration) was expected to take measures to address it. However, the report said that it was more important for the administration to help the disaster-affected people to reconstruct their lives and to maintain their health than to just focus on preventing cases of *kodokushi*.

The report argues that the characteristic that made the community unique is the people's shared experience of loss caused by the earthquake. It also points out that residents came from various other places and their new community in temporary housing was intended to be only temporary.

The report highlights the high proportion of elderly people at temporary housing. It divides temporary housing residents into three categories. The 'coping' group included those who have accessed professional support for health and welfare. The 'potential' group included those who did not currently need professional support but they might need it in the future. The 'healthy' group was comprised of those who were relatively healthy and who could expect a relatively easy life after reconstruction. The report said that community development in the temporary housing aimed to prevent the health level of the 'potential' group from deteriorating.

The research surveyed residents to identify those who needed care, and helped respondents to exchange information about public health activities. Public health activities provided counseling in addition to health care. The report also discussed community development and the importance of community events as a first step toward *renkei* (coordinated collaboration) and the development of a self-governing body. The report emphasized the importance of information exchange between *gyōsei* and volunteers, and between volunteers, in addition to other forms of collaboration among them.

The report of the Department of Health of the West Ward of Kobe City (1998) recorded the community development activities in the West Ward of Kobe City. The report, a consequence of the Hyogo Prefectural Nursing College report, presented action research focused on community development in Haruyama TSH from April 1995 to June 1997, after which the focus shifted to Natsu Aki PRH. It outlined the mental health of residents during these years. In 1995, the problems were insomnia, anxiety, difficulty in maintaining privacy and dealing with the noise of neighbors and loneliness, especially due to relocation to a new and unfamiliar place.

In 1996, it emphasized both the good and bad relationships between neighbors and the difficulty of establishing and maintaining relationships in temporary housing. It also stated that some people's health deteriorated due to the prolonged stay in temporary housing, especially due to the severe climate and temperature changes in temporary housing. It also noted the anxiety of people who were still left in temporary housing when people started to leave.

In 1997, the report focused on the worsening anxiety of those still left in temporary housing, while also noting the anxiety of those leaving temporary housing for permanent housing. This was not consistent with the media description of temporary housing as a better place than public reconstruction housing.

The report also described the activities of nurses in the role of life support advisor and health advisor, as well as those of volunteer nurses. It emphasized the need to develop a community and build up the support system for elderly and disabled residents in light of health issues. It observed that short-term community activities would not necessarily have lasting effects, so there was a need to think long term.

It observed that volunteers were important for the development of community at Haruyama TSH. It also found that those volunteers who were still involved after the first three years were residents of local neighborhoods and that this demonstrated the importance of the participation of local neighbors in order to establish community systems in temporary housing. However, it found that public health nurses and nurse volunteers could not cope with health problems in the community, recommending that more organized activities be based on public health centers in order to maintain the health of TSH residents.

Like the nursing college report, the Department of Health emphasized the importance of *renkei* between the public administration, especially the Department of Health, the welfare commissioner, the self-governing body, local neighborhoods and volunteers conducting home visits and organizing community activities. It suggested that the experiences of community development in post-earthquake housing should be seen in the context of broader issues relating to Japan's ageing society.

## Summary and conclusion

This chapter explained the process of allocation of places for TSH, and how this concentrated low-income class and the elderly people together, bringing them to the media's attention. People moved from evacuation centers to TSH several months after the earthquake. Those who could not find housing themselves and had no family moved to TSH. A large proportion of them were poor elderly people. According to Japanese values of fairness, allocation of TSH was made by lottery while giving priority to older people.

Elderly people were the major focus of allocations in 1998. While they remained a focus in 1999, the emphasis shifted to other groups, such as middle-aged people who were struggling and children, especially those made orphans by the earthquake. There were also efforts to address how to prepare for future disasters by looking ahead to the fifth anniversary of the earthquake, rather than looking critically at the past. As chapter three showed, in January 1999, the fourth anniversary of the earthquake, the total time spent on reporting about elderly people (550 minutes) made up 33% of the total time covering earthquake-related topics (1,675 minutes). In January 2000, the fifth anniversary, this percentage had dropped to 8.4% (200 out of 2,335 minutes), while the total amount of time spent covering the earthquake was greater.

The tone of media was analyzed from a gender perspective. Elderly women who lived alone were reported to have coped well and they were good at making friends, but this was more difficult for elderly men who lived alone.

According to the media, in 1998 PRH were mostly unhappy places; TSH were better in terms of personal relationships and human contacts, despite the poorer housing conditions. In January 2000, the media started to encourage people in PRH to look ahead to the future while maintaining a generally negative tone.

The media was critical of the public administration in many ways, if not all the time. But it is to an extent the nature of the media to focus on problems and be critical. It may be necessary to consciously look at the positive side before making a judgment and to pursue a balanced view of the real world.

Professional studies had a number of different viewpoints though they shared some similar views. Hirayama (2000) pointed out that the post-earthquake recovery process deepened the socio-spatial polarization of low-income and elderly people, and divided the city by concentrating victims and creating a group in need of welfare, a high proportion of which were those with low incomes and the elderly.

Two special reports on community care at TSH and PRH were prepared by the Department of Health and Welfare of Kobe City West Ward. The TSH and PRH were selected as the research sites in 1996 and 1998. Both reports presented data by age and drew attention to elderly people and those living alone. The Department of Health noted that the anxiety of those who were still left at TSH was a mental health concern.

Many people who left TSH had only good memories, but TSH was only temporary housing and had many problems. What made TSH life better was not the architecture but other things like receiving attention and free help from volunteers. The media reported that life in TSH worsened people's health, but at TSH I heard a different story.

It was not that people did not want to move to PRH; it was more a case of the anxiety caused by thoughts of leaving their community and moving to a new place. However, many were actually happy to be given a subsidy for a nice apartment with such low rent. Subsidized PRH and free TSH were important to residents who had limited personal finance, but the media did not talk about this aspect. Even with heavy subsidies, people who stayed at TSH until the end were concerned that they would not be able to afford the minimum rent for PRH.

The media's stories did not necessarily reflect what I heard from people. While it might be true that PRH residents would miss TSH life, it should not necessarily mean that they were not happy about their new life in better housing. The degree of happiness in both TSH and PRH rested on housing type and community. During a Mainichi TV broadcast on 14 January 1999, Prof. Sachiomi Kishimoto of the Osaka Education University said, 'People in public reconstruction housing do not complain much about the new housing. (only 18% in his survey complained). But they complain of the worsening opportunities for relationships with their neighbors and acquaintances.'

TSH had both negative and positive impacts on the health of residents. TSH was intended to be used only for one year, but many of the residents lived in it for three to five years. The buildings did not remain in good condition for long and eventually caused health problems. When not well built, they were too hot in summer and too cold in winter. Another issue of concern was the remoteness of TSH. Although this was first reported as an inconvenience, residents later reported an improvement in their health because they lived in clean mountain air and had to walk a long way to get to public transportation. This was good exercise for them and stimulated their circulation systems. Volunteers came in to help the residents in the post-earthquake community. With the encampment type housing, it was easier for volunteers to visit each household.

What made TSH life better was not simply architecture but other things like the attention and free assistance provided by volunteers. Although the TSH design was recognized as easier for community development, there were complaints about the design in terms of lack of privacy.

PRH residents expressed to the media their wish to return to community life in TSH. Some people still at TSH heard about how their former neighbors that had moved to PRH missed TSH. But my fieldwork suggested that it was not that people did not want to move to PRH; more so they were anxious about moving to a new place. In my fieldwork, many were actually happy to be given a permanent place to live that was affordable and subsidized by the government. Affordability was an important issue for people in my fieldwork.

The media representation of TSH life was not consistent with the observations I made in the field. It was slow to look ahead for PRH. Some people at TSH were worried about whether they would be able to pay the rent when they moved to PRH. Some people in general housing complained that those people who had been living at TSH for five years were not paying any rent and were occupying land for free. Some people at PRH said that people in the neighborhood did not talk to them because they knew that they were living in new apartment buildings for minimum rent. The rent was determined by the income of the resident. Some PRH residents felt very lucky that they were able to live in PRH and seemed like they did not want to be asked how much they were paying.

The majority of PRH residents had moved from TSH. In the next chapter, the stories of residents in PRH will be examined.

# Chapter Six
# Public Reconstruction Housing

## Introduction

A typical post-earthquake Public Reconstruction Housing (PRH) site is comprised of a group of modern high-rise apartment buildings. They have a computerized emergency call system and no common space or facilities in the buildings. There is one assembly hall for the use of all residents, but it is not big enough to include all the residents and is located outside the buildings.

The needs of elderly people were taken into consideration in the design of PRH, although PRH in general is meant for all ages. Housing built specially for elderly people is a relatively new phenomenon. The Silver Housing program provided daily support for elderly people through specially designed dwellings and care by life support advisors (LSAs). One LSA was in charge of as many as 60–100 elderly people, so they could not make daily visits. The Silver Housing[1] program was launched as a joint policy effort by the Ministry of Construction (MOC) and the Ministry of Health and Welfare (MOHW) in the early 1980s. It was planned by the Housing Department and the Welfare Department of the local government.

Derived from sheltered housing in the United Kingdom, Silver Housing is rented public housing that gives physically independent elderly people a barrier-free living space with care services such as a LSA and day care center. Silver Housing is a costly project that serves a small group of elderly people. Despite the initiatives of the central government in 1980s and project approvals by various local authorities in the following years, it remains a low priority due to the limited resources of many local authorities (Izuhara, 2000a: 131).

## Media attitudes to PRH

### *Reports on elderly people*

The media has drawn attention to the need to protect the health of elderly people living alone. It has predicted that 30% of the elderly will be living alone in Kobe PRH and, because of the ageing society, in Japan as a whole in

Photos 6-1: High-rise modern apartment building in the Natsu Aki PRH compound

the future. The central and local governments cannot protect senior citizens living alone by merely building housing for them.

In January 1999, Yomiuri TV interviewed an elderly woman who had moved into PRH:

> Six months ago I moved to this public reconstruction housing, but I have made no friends here. I do not feel positive any more. I feel a lot of anxiety. A day goes without any chance to talk with anyone. I watch TV all day and just sleep.

Neighborhood and community support networks will be discussed further in this chapter as well as in chapter seven.

Media reports do not necessarily represent the perceptions of elderly people. The media were slow to reflect changes in the perceptions of some elderly people. The media focused on negative aspects of PRH in 1999, but became more positive in 2000 as it looked to the future. The media tone in 1999 did not necessarily reflect the feelings of elderly people.

Another Yomiuri TV program, broadcast on 10 January 1999, reported the case of Mr Hiroshige Takehara (71). For him, life in TSH had affected his *mushibamu* (health). Similarly, a NHK program, broadcast 14 January

Photo 6-2: Ground floor apartment for a disabled person in the Natsu Aki PRH complex. The slope for the wheel chair is at the back of the veranda.

1999, reported that Mr Yasuo Uematsu (70) lived in TSH for two years. A draught there worsened his asthma.

The media reports in 1999 were very biased towards a negative perspective. On the other hand, the media adopted a more positive tone in 2000. The year 2000 was both the fifth anniversary of the earthquake and the turn of the century. The media's tone changed partly because the fifth year anniversary in 2000 was the *kugiri* (separation time) and *fushime* (bamboo joint); it was time to start a new outlook. The media decided at this point that its policy was to look forward in a positive way.

### *Tone of media assessment of PRH*

The *tetsu no tobira* (iron door), a door used in modern apartments instead of the wooden doors used in traditional Japanese houses, often featured in reporting on the isolated living arrangements in high-rise PRH.

According to the media, the high-rise apartment buildings were a failure. A report on Ms Nakagawa (85) on Yomiuri TV, 24 November 1998, said that a 'problem of living in the modern high-rise building was that an old woman lived alone. She was alone for whole three months'. This news program blamed the *gyōsei* (public administration) for the situation and called on them to take more responsibility.

The Governor of Hyogo Prefecture praised in a TV program the Department of Construction for constructing high-rise apartment buildings so that they could meet the need for PRH. Life in a high-rise apartment building may not have the features that one-story tenement housing has, but there were few other options for meeting the required number of units.

Another TV program saw the process of *seikatsu fukkō* (life reconstruction) in TSH as a success. It asked why the developers of PRH have not learnt from the experience of TSH.

## Fieldwork in PRH communities

### *Background*

As explained in chapters two and four, the main part of my fieldwork took place at the Natsu Aki PRH, the Fuyuyama PRH and the Haruyama TSH (which is near the Natsu Aki PRH). The Natsu Aki PRH and the Haruyama TSH are in the West Ward of Kobe City. The Natsu Aki PRH comprises the Natsuyama PRH and the Akiyama PRH. The Fuyuyama PRH is in the Central Ward of Kobe City. I also visited three other PRH sites and two other TSH sites which I used to make comparisons with.

The community care dynamics were different between the two PRH communities where I conducted my fieldwork. While similar actors could be identified in both communities, such as newly created public staff like the LSA (Life Support Advisor) for elderly people living alone, the major difference between the two was in their source of initiative and leadership. The large suburban community had active welfare commissioners and volunteer groups from the neighboring community who had moved to the area before the earthquake as the planned new town developed. The smaller section of three buildings had a self-governing body which included a chief who was an active leader. The other nine buildings were having problems establishing their self-governing bodies and no chief had emerged to act as the official representative. Although the chief was officially representing the two main buildings, he was also taking care of the Silver Housing scheme because they had not established their own self-governing body.

### *Natsu Aki PRH*

Natsu Aki PRH is a city government public housing scheme in a suburban new town in the West Ward of Kobe City. The new town started to develop on the east side of Natsu Aki station two years before the earthquake.

The central area of the town is situated around the station. It has basic town facilities including shops, a supermarket, a bank, a post-office and a town community center. There are various housing types, including houses and apartment buildings. More affluent residents live on the east side. They started to move into this area a few years before the earthquake. I learned during my research that the east side was designed and built as a dream new town for the middle-class based on the model of Senri New Town in Osaka. It was only two years after the east side was built that the earthquake occurred. The east side was still developing its own community. After the earthquake, 1,500 subsidy dependent households were brought in without consultation with the neighborhood.

The west side had three kinds of areas: two groups of PRH at Natsu Aki (Natsuyama PRH and Akiyama PRH) and one group of *kōdan* (public corporation apartments). The Natsu Aki PRH compound had three buildings of 500 households in total. The Akiyama PRH compound, which was just next to the Natsuyama compound, had 11 buildings for 700 households in total. Haruyama TSH compounds were located next to the new companies and factories in the mountains to the west of the east side of the new town.

Those on the east side bought expensive but nice new residences and moved in a few years before the earthquake. Several months after the earthquake, several thousand lower income people moved to TSH compounds nearby and then three to five years after the earthquake into the heavily subsidized modern high-rise apartment buildings on the west side. Their monthly rent was as low as JPY 6,000 (£30).

The TSH was meant to be open for one or two years but it took five years for the Fuyuyama TSH compound to be closed down and removed. Some people moved to other PRH compounds in the West Ward or to other wards such as Nagata, Nada, Higashi-nada, Hyogo, Tarumi, Suma and Chuō (Central Ward). Some moved to the Natsu Aki PRH compound. These areas are administratively within the same district of the West Ward of Kobe City.

On the west side, the residents in the Akiyama PRH moved in earlier than the residents in the Natsuyama PRH by more than two years. They were family groups and younger people who really wanted to leave TSH. In Natsuyama PRH most residents were elderly people living alone. There were also many disabled people among them. They were late comers to the Natsu Aki area in 1998 because they had not won the lottery for other PRH in urban areas and few were eager to move.

The Akiyama PRH was designed for families, so had bigger apartments with more rooms. The Natsuyama PRH had smaller apartments, often only a single room, and was designed for those living alone.

The Natsuytama PRH had four different apartment types: Silver Housing, apartments for couples, apartments for disabled people and apartments for single people. Silver Housing is rented housing for people aged 65 and over or couples that have at least one person aged 65 or over. The apartments are linked to a special old people's (*tokuyō, tokubetsu tōgo rōjin*) home and applicants to Silver Housing need to meet certain requirements. Silver Housing is designed for barrier-free living; there are no steps and, rather than Western-style push and pull doors, sliding doors make the apartment more accessible for wheelchairs. The apartments have hand rails on the wall for safety and an emergency call button links the resident to the LSA office[2] (Wilson, 2000: 148). The apartments have only two rooms, a dining-kitchen and living-bedroom. One version has the kitchen closer to the entrance while the other has the kitchen towards the back.

Kobe City has been trying to introduce a new system and have made a special fund available to subsidize the expansion of staff numbers, including increasing the numbers of LSAs and Life Reconstruction Advisors for Older People. Among the staff, LSAs are the most visible because they are numerous, receive frequent media coverage and have an office on site. Local government dispatches LSAs from the Social Welfare Corporation (*Shakai Fukushi Hōjin*). They make regular home visits and give daily advice to people aged over 65 and living alone and to residents of Silver Housing.

Apartments for couples feature an entrance space, a kitchen dining room and three bedrooms. The kitchen is wider than in Silver Housing. The apartments are also equipped with an emergency call button but receive fewer services than Silver Housing.

The apartments for the disabled are on the ground floor. At the back of the residence, there is a slope for wheelchair access via the back balcony. The entrance has a sliding door and, like Silver Housing, these apartments are designed to be barrier free. Beyond the entrance space, there is a main room with a kitchen. There is a tatami mat room with a *kotatsu* (Japanese winter table with foot warmer). The floor in the tatami room is higher than the rest of the apartment by about half a meter so that the resident can easily move from a wheelchair in to the room. Beyond the tatami room is a bedroom, which is the same height as the main room. There is also a bathroom.

Apartments for people living alone are on the higher floors. These are not designed with the barrier-free concept but they have more emergency call buttons than other family apartments. On the left hand side of the entrance is a small room which has several closets. Towards the back is a kitchen and dining room, on the left of which is a room with *kotatsu*.

## *Fuyuyama PRH*

The Fuyuyama PRH was built in the *Chuō* (Central Ward), the central part of Kobe City. The compound was built on land where a private girls school had been, next to the Nakayama Catholic Church and a couple of old city-run apartment buildings. It was much smaller than the Natsu Aki PRH with only two high-rise apartment buildings and a community center. The apartment buildings were for *dōtai*, a special program promoting *dōwa*.[3] Enomoto writes,

> The compound was built in April 1998. It has 550 households accommodating 839 people in total. Of these, 220 households are silver housing designed for elderly people living alone. These households are served by four LSAs.[4] LSAs offer the following services: overseeing residents' safety; responding to emergencies; temporary assistance in housework; counseling and guidance on daily life activities; offering various information; liaison and coordination with related agencies; and all other services and assistance required in daily life. An assistant offers such services during holidays and at night etc (1996:144–145).

LSA services were available to residents aged over 65 who did not live in Silver Housing apartments. To move into a Silver Housing apartment, a person world need to apply for it and be selected.

## *Activities*

Natsu Aki PRH and Fuyuyama PRH differed in their size, key actors and activities. The events and actors are summarized in Tables 6.1 and 6.2. The main activities at the Natsu Aki PRH were the lunch program, mini-day care service, health promotions, tea parties, the *yūai* home visits, ground golf, BBQs and the summer festival. The main activities at Fuyuyama PRH were the breakfast service, health promotions, handicraft service, karaoke and the summer festival.

### The lunch program

The lunch program took place twice a week at the Natsuyama PRH. This was a simpler activity than the others. Residents would come out from their apartments to have lunch with their neighbors and to exchange information in a warm and friendly atmosphere. The atmosphere was one of the measures used by an organizer to see if the activity was successful. A harmonious and friendly atmosphere with no confrontations, fights, or arguments was favored. It was important that everyone enjoyed the activity and left with a good feeling.

Table 6-1: Activities at the Natsu Aki PRH

| Purpose | Who runs it | Target population | Place | Frequency |
|---|---|---|---|---|
| *Lunch service* | | | | |
| Eat with others<br>Going out<br>Meet people<br>Balanced diet | Welfare commissioner<br>Life reconstruction advisor for older people<br>Volunteer group from the East side | Elderly people living alone<br>About 20 people | Akiyama community center | Twice weekly |
| *Mini day care service: physical exercise; entertainment; lunch; game; afternoon tea* | | | | |
| Relieve carer's burden<br>Meet people | Welfare commissioner<br>Life reconstruction advisor for older people<br>Nurses from the Dept. of Health of the West Ward<br>City official<br>Volunteer group from the east side | Disabled old people and their carers<br>About 20 people | Akiyama community center | Once a month |
| *Health promotion and tea party: health counseling; blood-pressure measuring* | | | | |
| Health promotion<br>Meet people<br>Info exchange | Dept. of Health<br>Welfare commissioners<br>Volunteer group<br>Public health nurses | Residents<br>Over 50 people | Natsuyama and Akiyama community centers<br>Kōdan Aki | Once a month at each community center |
| *Monthly birthday party* | | | | |
| Celebrate birthday for the month | Volunteers | Residents aged over 65 living alone | Natsuyama community center | Once a month |

| | | | | |
|---|---|---|---|---|
| Yūai *Home visit* | | | | |
| To make sure an old person living alone is doing well<br>To initiate communication among new neighbors | Welfare commissioner<br>Volunteer groups from the West side | Requested in *Yūai* visit survey by elderly people living alone<br>Twenty people; ten for each team. | Entrance of each apartment | Once in two weeks |
| *Sport (ground golf)* | | | | |
| Healthy physical exercise<br>Enjoyment with friends | Initially organized by Public Administration and welfare commissioner) with assistance of the East side old people's club<br>Now run by themselves | Old residents of Natsuyama.<br>About 40 to 50 people.<br>Separate group for the disabled people. | Ground park next to the compound | Once a month |
| *BBQ* | | | | Twice a year |
| *Summer festival* | | | | Once a year |

Table 6-2: Activities at the Fuyuyama PRH

| | Purpose | Who runs | For whom | Place | Frequency |
|---|---|---|---|---|---|
| Breakfast service | Meet people in the community | Self-governing body volunteers. Catholic Church nuns and volunteers, 12 in total. | Residents | The Catholic Church dinning hall | Once a week |
| Health promotion | Promote healthy life | Dept. of Health, public health nurses, nutritionists, a doctor | Old people living alone | Community center | Once every two months |
| *Yuai* home visit | Make sure that old people living along are doing well | Life Support Advisor (LSA) | Old people living alone | Entrance of each apartment | |
| Handicraft Class | Enjoyment | Volunteers from the Sōkagakkai religious group | | Community Center | |
| Karaoke | Enjoyment | Residents | Residents | Community Center | |
| Morning Radio Exercise | Healthy life | *Jichikai* (self-governing body) | Residents | Ground in front of community center | Every morning 6:30 am |
| Festival | Activate community Enjoyment | *Jichikai* (self -governing body) | Residents | Ground in front of community center | Once every few months |

There was a charge for the lunch box, but tea was provided for free. The purpose of the lunch service was not only to provide a balanced diet, but also to provide human contact. The ultimate goal of all the community activities was to provide an opportunity for residents to make friends. Volunteers would serve lunch, sit and talk with people and facilitate communication with other residents. Not everyone was social; some residents would participate in the lunch service but sit alone without talking to anyone. Some waited to be talked to. For example, one man aged over 70 that I spoke to complained that the lunch service was the same as eating alone:

> Once I went to your lunch service program. I thought I could talk to other people, but it was just eating lunch quickly and everyone left quickly. There was no reason for me to go. I had expected that we could talk for one hour or two.

**Tea party**

The tea party (*sawa-kai*) was a major event in TSH communities. Residents expected similar opportunities to meet people and make friends as they had at TSH when they moved to PRH.

**Mini-day care service**

At Natsu Aki PRH, Mrs Sakura and Ms Ume, Life Reconstruction Advisors for Older People, said on different occasions that it was difficult to develop a mini-day care service because of the competition they faced from private professional day care services. The day care service run by the old people's home had a bathing service. The bathing service is one of the most important of the services because bathing is valued in Japanese culture. If the schedules for private and community services conflicted, the elderly residents would choose the service offering bathing, even though they have to travel to the bath at the old people's home. This is an example of the problems that face small, charitable programs (Wilson, 2000: 133); they need monopoly in order to secure participation of the residents.

Figure 6.1 is a map of the PRH area on 1 September 1998 (Heisei 10 according to the Japanese system) with demographic information on the number of households and the total population in the area. It says, 'Human to human, heart to heart. Let's build a welfare town for our future.' and lists community activities such as the tea party and *yuai* visit, the health exercise meeting for rehabilitation, *mimamori* and activity coordination meetings and ground golf. It indicates that the PRH employed twenty staff: two males and eighteen females.

Figure 6-1: Volunteer's newspaper with community map

ボランティアいぶき新聞 3号

発行人：ボランティアいぶき 代表者：坂本 津留代

**Human to human. Heart to heart.**

**Let's build a welfare town for our future!**

Current activities: Tea party, Yuai home visit; Health promotion programme with rehabilitation; Community meetings; Ground golf.

Staff: Male (2); Female (18).

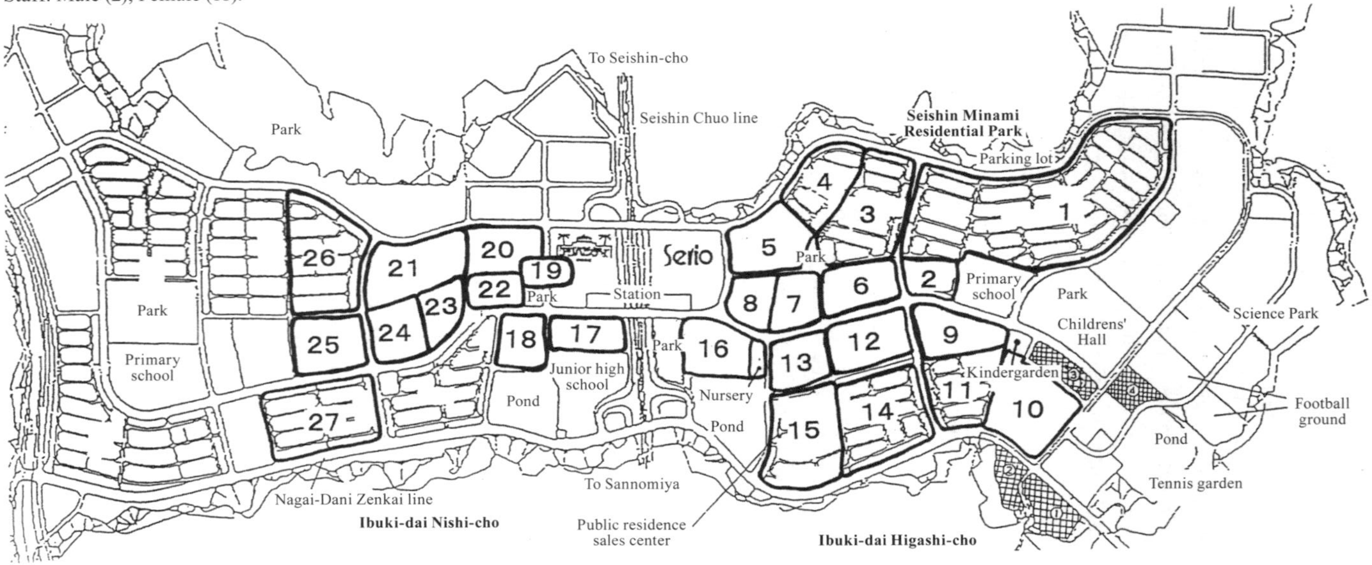

Demographic data of the new town (as of 1 September 1998)

Population: 15,323 people, 5,028 households

| No. | Name & Number of households | | No. | Name & Number of households | |
|---|---|---|---|---|---|
| 1 | 5 chome individual housing | 512 | | | |
| 2 | Harmony Town | 51 | | | |
| 3 | Garden House | 148 | | | |
| 4 | 3 apartments | 108 | 17 | Earth City | 256 |
| 5 | Grand Colina | 405 | 18 | Louise Fille | 252 |
| 6 | Belle Air | 171 | 19 | West City Court | 165 |
| 7 | Vios | 249 | 20 | City public housing No.1 | 456 |
| 8 | Ellio | 218 | 21 | City public housing No.2 | 690 |
| 9 | Villa Cousille | 70 | 22 | Floure | 216 |
| 10 | East City Court | 528 | 23 | Ville de vin | 237 |
| 11 | 4 chome individual housing | 100 | 24 | Hills | 296 |
| 12 | Mont Seine | 201 | 25 | A Lovely | 222 |
| 13 | Southern Court | 95 | 26 | Individual housing | 166 |
| 14 | 3 chome individual housing | 114 | 27 | Individual housing | 160 |
| 15 | 4,6,7 apartments | 154 | | | |
| 16 | Southern Hills | 384 | | | |

Note: Numbers 1–16: Higashi-cho (Eastside town). Numbers 17–27: Nishi-cho (Westside town).

## Home visits

In the *yūai* (friendship and love) home visit program, volunteers visited elderly people living alone to talk to them and to see if they were well. At the Fuyuyama PRH *yūai* was done by LSAs. At the Natsu Aki PRH, a volunteer group was established at the initiation of the *minsei iin* (welfare commissioner) because LSAs had not been visiting systematically, partly because the visits were difficult for an inexperienced young person (Keith, 1982, Wilson, 2000: 7) and partly because there were too many residents to cover. Mrs Sakura said that it took six months for volunteers to be accepted by the residents. When a visitor knocked on the door of a home, the resident often answered, 'I am fine,' and did not open the door. It was not like in an evacuation center or TSH where residents tended to be more eager to exchange information about disaster relief and so on.

A visitor cannot just walk into a person's residence. In some cases pushy salesmen selling expensive things and targeting elderly people living alone were reported as problems. This is not a violent crime like in the Chicago case (see chapter two), but it still makes residents worry and has implications for the sense of the community security. While elderly people are cautious not to answer the door to such salesmen, they will open it to a welfare commissioner because the status of welfare commissioner is trusted in Japanese society.

In the *yūai* visit, it was not unusual to find someone who had not talked to anyone all day; some said they had not talked with anyone for three days. For some, their interactions were just formal, superficial exchanges; they might not feel that they had developed relationships or made friends. This shows the importance of providing an opportunity for human contact. The *yūai* visit might bring human contact but it is not enough.

Mrs Sakura aimed to gain the trust of residents gradually. During her first visit, she would stay for only a short time for *kaoawase* (showing her face). Other activities such as the lunch service program were needed to create such an opportunity for human contact. Mrs Sakura saw this as a starting point for community generation, but she aimed at more than this. Rather than just having staff visit the residents, she wanted to see the elderly residents 'come out' to do things themselves.

## Outdoor activities

Outdoor activities like BBQs, ground golf, and festivals were organized. According to Mr Takada, ground golf at the Natsu Aki PRH compound was started by *minsei iin* who, with the help of the *gyōsei*, borrowed equipment and encouraged interested residents to organize a club. The aim was to begin by running games and then form a club if that was appropriate for the needs of the community. Another club was formed for disabled players and both groups are now independent. Over time, the ground golf club might become a more general club for elderly people.

Most activities focused on elderly residents with priority given to those living alone. This was the case with all the regular activities. PRH housing was not designed for elderly people but for the poorer people who could not afford to rebuild their homes after the earthquake. A high proportion of the residents of high-rise apartment buildings were elderly, with a high proportion of these living alone. This situation can be said to be a forecast of what is to come in Japan's ageing society.

Although most activities were targeted at elderly people, Mrs Sakura opposed the tendency to see elderly people who lived alone as needing special attention. Younger couples or families with children have different problems of their own. She drew attention to people in their 40s and 50s and young people who actually needed more support to get their life going. The report on Haruyama and Natsu Aki Healthy Community Generation by Kobe City West Ward Department of Health argued that festivals were important for community development among these groups. Different groups need to work together in preparation for a festival. According to Cockerham and Yamori (2001), summer festivals are key to longevity in Japanese society.[5]

In addition to these activities, *minsei iin* conducted a household survey at the Natsu Aki PRH at the request of Kobe City in 1999. LSAs for elderly people and volunteers helped with the survey. The planning for all activities was based on the results of the survey and elderly people, especially those living alone, were invited to participate. Those who were visited by the *yūai* had indicated in the survey that they wanted such visits. Despite the fact that participants were selected on the basis of their survey responses, one might still question the selection process given the small number of participants. At the time of my fieldwork, the survey had not been completed.

Having a recreation room for common use is important in community generation. A common room is where people gather to begin an activity. The community assembly hall was too small to prepare seats even for 100 people so was not large enough for the compound. Initial activities were small-scale and programs were expanded by a 'think-as-you-go' approach. The space issue was something that had yet to be resolved at the time of my research.

### *Coalitions*

An important event at the Natsu Aki PRH is the regular *mimaomori* (watch with care) meeting. In an NHK earthquake special, 16 January 2000, experts reported that many *machizukuri kyōgikai* (town development committees) had been formed in Kobe after the earthquake. The chair was Mr Tamura of NHK Kobe and the experts were the Hyogo Prefecture Governor Mr Kaihara, an economist Mr Kazuto Uchihashi, a nurse and representative of the *shimin kikin* (citizen fund) in Kobe, Ms Hiroko Shirono and Mr Yoshiro Fujiyoshi, an NHK editorial writer. The experts agreed that the formation of 150 *machizukuri kyōgikai* was a symbolic milestone in the post-earthquake community generation process. The *mimamori* meeting at the Natsu Aki PRH was one of the functions of *machizukuri kyōgikai.*

During my field work, attendants of *mimamori* meetings at Akiyama PRH were welfare commissioners, volunteer groups and the *jichikaicho* (chief of the self-governing body). At Natsuyama PRH, attendants were *yakuin* (committee members), Kobe City officials and life reconstruction advisors.

During meetings, those in charge of activities reported on how the activities of the past month had gone and discussed problems in the community. If an organization wanted to come into the community to run an activity, they had to present the aims and purpose of the activity at the meetings. The meetings enhanced *renkei* (coordinated collaboration), a key requirement for community generation.

The US Centers for Disease Control and Prevention (CDC) (1997) has a section on Coalitions:

> Engaging the community will very often involve building coalitions of diverse organizations. A community coalition can be defined as 'a formal alliance of organizations, groups, and agencies that come together to work for a common goal' (Florin et al., 1993, p.417). Coalitions are usually characterized as 'formal, multi-purpose, and long-term alliances' that 'fulfil planning, coordinating and advocacy functions for their communities' (Butterfoss et al., 1993: 316, 318). They can be helpful in a number of ways, including maximizing the influence of individuals and organizations, exploiting new resources, and reducing duplication of effort. While the literature reveals that coalitions have not been systematically studied and contain little data to support their effectiveness, funding sources have been giving serious commitment to developing coalitions as an intervention to address health issues (Butterfoss et al., 1993).

## Analysis of community work

### *Leadership*

Although both the PRH compounds had *jichikaicho* and LSAs, they had different kinds of volunteer groups and welfare commissioners. Leadership also differed between the two PRH compounds. The Fuyuyama PRH was lead by one man, Mr Tanaka. The leadership of the Natsu Aki PRH was shared but the key person was Mrs Sakura.

Mrs Sakura had multiple official roles. She was an active *minsei iin*[6] of the east side and the vice chief of the welfare commissioners. She also led the volunteer groups from the neighboring community on the east side. She was the chief of the self-governing body of the east side and the chief of the united self-governing body of the district. None of these roles were paid work, but they came with honor status in Japanese society. *Minsei iin* has a different meaning to 'volunteer' in the Western sense; *minsei iin* work with the municipal government and are from the community.

Mrs Sakura had a clear set of goals for the PRH community in her mind. She had a philosophy about long-term aims, how to develop community and how to maintain community once it is developed. She took a bottom-up approach to leadership, compared to Mr Tanaka, who took a more top-down approach. For example, Mrs Sakura liked people to come voluntarily to an activity like the handcraft class. She would say, 'I do not oppose loudly but I do not agree with an approach that says "Do this, do that". I am very sad about such approaches'. Mr Tanaka, on the other hand, would say things like 'I need to teach my people. I need to train them' and 'All the effort I put into training at the temporary shelter, now I have to do it again here at the permanent housing and it is really challenging'. When Mr Tanaka said 'train', it was like he was from the older generation, like he was talking about soldiers or unskilled employees. However, this was not school education but local community. Although he also drew on democratic approaches and understood their importance, he acted like a one-man leader. In an interview in the 10 February 1999, an accountant of the self-governing body of the Fuyuyama PHR said,

> Mr Tanaka knows everyone related to the *gyōsei*. It is not that good. I am not saying that I doubt him, but there should be witnesses. He should bring other witnesses for meetings to observe. He is old. Nobody can listen to any one talk one hundred per cent. He does not record the talk. It is better to go with two or three people to discuss an important issue. He is a man with a sense of responsibility. He is a bit pushy, though. But he is a man of fairness. I like that in him.

Mr Tanaka went to meetings with *gyōsei* alone and did not bring other *yakuin*s (committee members of the self-governing body) along. Also, the informant's comment that Mr Tanaka was a bit pushy indicates that Mr Tanaka tended to adopt a one-man leadership style based on his understanding of what needed to be done. However, he was not a dictator. The *yakuin* saw that he worked with a sense of responsibility and fairness. For example, when challenged by an angry resident on the street, he would patiently listen to their complaint and encourage them to bring up their concerns in a formal meeting.

Mrs Sakura always talked to people. She would say, ‘Let’s do something together’. She talked about her ideas but did not tell people what to do. She would give examples of how people could help each other in daily life and would propose ideas for community activities and events. She would facilitate concrete actions. For example, she said, ‘Shall we do *hanami* (a picnic during the blooming of cherry blossoms) in April? It is very different eating outside from eating here inside. It will be a good feeling to eat on the grass outside when it gets warmer’. Would residents want to do these things if they were told to by a more dominant leader?

Mrs Sakura also wanted to develop the community so that she could spend her old age there comfortably. She was conscious of the ageing society. In an interview on the 16 February 1999, she said,

> This is a model of an ageing society. I am sorry to say this, but I think this is a test. We are experimenting on them. If it does not work here, how can we cope with the ageing society that we will all be facing soon? This place is far ahead than the rest of Japan in terms of the ageing society. 300 households here are aged households. I think if things go well here, then Japan’s ageing future will go well. This ageing will happen within ten years.

From a governance perspective, her leadership had a bottom-up approach and was based in the community. From a community perspective, this might not have been so as she was chair of everything.

Mrs Sakura distinguished between ‘outsider’ volunteers and ‘insider’ volunteers. She was resistant to ‘outsider’ volunteer groups working in the community. In this respect, she might have appeared controlling. She was much happier when volunteers from the west side Natsu Aki began the *yūai* visit in February 1999.

She was also cautious about professional care workers operating in the community. Although large PRHs needed more community service centers and for-profit organizations have better resources, professional services would limit the opportunities for spontaneous community development. Community programs lacked competitive resources and, unlike professional organizations, had no start-up finance.

Immediately after the earthquake, 'volunteerism' was seen as a great new development in Japanese society. One of the good outcomes of the disaster was that people began to work as volunteers and to help others in need. However, Mrs Sakura saw the volunteers who helped those at evacuation centers and in the TSH as a problem in the PRH context. Although volunteers helped when they could, they were outsiders who could leave whenever they wanted to without taking responsibility. They also made residents feel entitled to free help rather than encouraged them to be independent.

At the Natsu Aki PRH, 50 yen (about 25 British pence) was charged for a cup of tea. Some people who were used to receiving free tea and food complained; 'Are you saying you are collecting money from the poor earthquake-affected people?' Mr Tanaka shared the same view, 'Many volunteers came to help at the temporary shelters and that made people very dependent'.

Mrs Sakura believed that working for others rather than needing other's help is important for happiness. Asking other people for help makes a person indebted and puts them in the power of others. She said, 'I do not want to become a complainer when I get old'. This idea of happiness in relation to loneliness will be discussed in chapter seven, along with *tsunagari* (being related to someone).

A Kobe City official who would often help in the PRH said to Ms Ume, a Life Reconstruction Advisor for elderly people, 'There is a *tochi* (place, land) where it is easy to raise volunteers and generate community, and there is one where it is not as easy'. According to Ms Ume, the reason for this was, 'There are many people who want to do something but there are not many who take the leadership. This PRH community is more active but at the other PRH community where I am in charge nothing is happening'. Shc attributed the lack of initiative from the residents at the moribund community to a lack o lcadership.

Leadership with good coordination is critical for community generation. A leader needs to understand and hold together the group so that the community can develop (Vanier, 1989).

### Coordinated collaboration

*Renkei* (coordinated collaboration) works well when key actors take multiple roles. In my fieldwork, *renkei* was achieved in two ways, by good communication between representatives with responsibility and by having a key person taking multiple roles. This promoted better integration, collaboration, and coordination of resources from multiple community sectors.

For example, a LSA for elderly people who was helping with the survey and participating in the *yūai* home visits knew many things about individual residents that the lunch service volunteers did not. The chief of the lunch service volunteers thought that the LSA's participation in the lunch service program made a difference in stimulating people's conversation. This shows that coordinated collaboration and good communication are important.

Good communication is necessary to mobilize resources. At the Natsu Aki PRH, the health promotion program and tea party took place together. The welfare commissioners and volunteer groups organized the tea party in the same hall that a public health office employee went to for the health counseling program. People would talk with others over a cup of tea while waiting for their turn for health counseling. Many people just went for a cup of tea and did not seeking health counseling. Welfare commissioners coordinated with volunteer groups like Co-op and Yamabiko to join their activity. These were some of the same volunteer groups that worked at Haruyama TSH. The community and volunteer groups liaised with the Department of Health to make their work more effective.

At the Fuyuyama PRH, the chief of the self-governing body did *renkei* with *gyōsei*. In comparison, because the self-governing body had not yet been formed at the Natsu Aki PRH, people like Mrs Sakura were taking a key role in *renkei*, along with volunteer groups and *gyōsei*. The Natsu Aki PRH was relatively active thanks to the intervention of people like Mrs Sakura and volunteer groups from the east side, but there were limits to what they could do without self governing bodies of their own.

The Natsu Aki PRH needed a *jichikai* (self-governing body) to represent and take responsibility for the community. On one occasion, Mrs Sakura said that the lack of a self-governing body at the Natsuiyama PRH was not due to the age of the residents, but on another occasion she mentioned that age was a factor that made it more difficult. Other factors that she put forward were insufficient funding, the attitude of the residents and the size of the community. In an interview on the 8 July 1999, she said,

> I do not think it is a matter of age. They are all from temporary shelter housing. They lived like that for four years. They received support from the Kobe City government. They cannot longer stand on their own feet. They remember how things were at the temporary shelters. They know how to spend money given for free. They know how to get together and have a party with the subsidies they received. There are too many of the *yakuin*s (committee members).

She added,

> It is annoying that these *yakuins* become suspicious that we, the volunteers, are receiving money. We do not. This is no longer disaster relief life at an evacuation center or a temporary shelter. After all, it is a matter of money. I know money is not everything. Still, whether you have money or not makes a difference. They talk as if they were managing a place of 120 or 200, so it is different from the real situation here.

The old age of residents is not in itself the reason why it was difficult to form a self-governing body. The compound was large and built all at once, not gradually. It was not a small town which developed over time. It also lacked transparent communication and the leadership to manage communication to encourage the voices of the residents.

At a *mimamori* meeting on the 9 July 1999, Mrs Sakura said,

> You need to form a *nin-i dantai* (voluntary appearance group). Those who want to join do join, those who do not want to do not join: a *rengō* (union), any name, just name it. Invite everyone and make everyone listen. To those who misinterpret we cannot afford to pay attention. Ignore them. But try and invite them to one place. This is the key. If they talk separately, it will never finish. If one does not want to join, then go without them. If they join, then go with them. Then those who do not come also have a responsibility. They need to stand up and speak out themselves.
>
> If one starts to interrupt, fights can start. Have them finish first. Have them listen to the end first. Remember the *jichikai* (self-governing body) is not for *yakuin* (committee members). There is no need to restrict *yakuin*s, invite all those voluntary people. Then *jichikai* is almost formed, but then always it gets destroyed. Discussion is most important. Make them talk and listen.
>
> Anyway, you need to form a self-governing body. We need someone who is strong and can control. The person says, 'Listen!' If they knew how to form a *jichikai* (self-governing body), then they would not get into such trouble. Remember, there is not unanimity (*zen-in icchi*). Don't seek the impossible. You need to start from gathering together and then listen. Ignore those who do not come to the meeting.

As stated earlier, it is important that a *chiiki* (local community) has its own self-governing body, otherwise the *renkei* with *gyōsei* is difficult.

On 10 February 1999, a LSA at the Fuyuyama PRH compound said,

> I thank Mr Tanaka. He does very well. It is very difficult for us to just come in to organize a program like this one. Mr Tanaka first did *jinarashi* (preparation)

> here at the self-governing body. Without it, we cannot just come and do this. The first *pinarashi* is the difficult part. It will be too much for us to do *jinarashi. gyōsei* (public administration) alone cannot do it. There needs to be *ugoki* (movement and dynamics) of the community and local society. *Gyōsei* should support them.

On 26 January 1999, The Director of Department of Health and Welfare of Kobe City also said,

> A self-governing body is being established at every public reconstruction housing. This enhances the health and welfare project development.

This shows that messages from those that worked in the PRH reached the Kobe City level, although there was a gap between the perceptions at the community, Kobe City and Hyogo Prefecture levels, as well as at the central government level in Tokyo.[7]

*Renkei* includes working together. Researchers at the US Centers for Disease Control and Prevention (CDC) have demonstrated that a population can achieve long-term health improvements when people become involved in their community and work together to effect change (Hanson, 1988–89).

Using examples from the field of housing, health and social services, Heywood et al. (2002: 137–139) demonstrate that working together involves understanding, cooperation, collaboration, coordination, networks and partnerships, using the example from the area of housing, health and social services (Heywood et al., 2002: 137–139). Understanding is usually obtained through a care manager who has knowledge of housing. A form of collaboration is contact between housing wardens and health and social services on behalf of a resident. Collaboration involves professionals and agencies working together on joint projects, for example, to reduce the isolation of elderly residents. Coordination encourages various agencies to work together to achieve agreed objectives. Networks promote informal meetings where views are exchanged, mutual understanding is improved and cooperation and collaborations are developed; they can be a precursor to formal coordination arrangements. There is often an expectation from government that health and social services work in partnership.

### *Gender and community politics*

Leadership is the key to successful community development and those who take such key roles are not always men. Although often unrecognized. women also take on a vital role in their communities (Moser and Peakeeds, 1987). A male welfare commissioner at Natsu Aki PRH and Director of Department of Health and Welfare said in separate interviews that the power of women is a driving force in community generation (Dominelli, 1990). As presented in chapter four, although both elderly men and women feature on TV programs, reporters and commentators always say, without any exception, that men need to try as hard as women. Women's work was recognized.

However, there were gender issues. Women tended to be thought of as a group, unless men were working directly with them when they might be mentioned by their individual names. Levels of activity also tended to be different. Official representatives' roles tended to be taken by men. Community management roles and work were primarily undertaken by women at the community level, whereas formal community politics roles were undertaken primarily by men at the community level (Moser, 1993).

Mrs Sakura, the female leader at the Natsu Aki PRH compound, was taking both a community management role and a community politics role. In this way, she was one of the few exceptions. The official representatives of most groups were men, even when the real leadership and decision-making was taken by women. By taking multiple roles, Mrs Sakura coordinated all the groups well at both the public administration and the local levels, but the general view was that things would go more smoothly if there was a man in her role. Mrs Sakura and Ms Shirono, a nurse volunteer leader at a TSH where I visited but did not conduct fieldwork said that when they argued at meetings, men said, 'You are just a woman, shut-up' and 'Why is your district sending a female representative?' Both were highly respected in their community. At another PRH which I visited but did not conduct my main fieldwork, the chief of the self-governing body was a man but everyone knew that the vice chief, a woman, was the one making all the decisions. The belief was that any group should be headed by a man and women should provide invisible support. This belief was particularly strong among elderly people.

At the Natsu Aki PRH compound, Mrs Sakura knew the resources and limits of the public administration. She had ideas about collaboration and understood that women had a bottom-up role in policy, allowing policy to directly reflect needs at local level. Mrs Sakura talked to the residents directly on a daily basis, encouraging them to start community activities and establish a self-governing body. She was available and gained the trust of residents, helping to solve problems when they arose. She was not working for her own personal interests, but considered what was important in the long run and sought ways for the members of the community to be independent.

Although, in Mrs Sakura's case, community development was very successful, this does not mean it would be a failure if a man was in control.

## Perspectives on media attitudes

In my fieldwork, people occasionally talked about the way that TV stations reported on life in TSH and PRH. The media is no doubt an important information source, but people do not necessarily trust or agree with what it says.

The change of the media to a more positive tone can help to explain the more positive attitudes I observed during my fieldwork. During my field visit to the PRH in the summer of 1999, the *jichikaicho* (chief of the self-governing body), Mr Kamiyama, said, 'Our goal is that those who had such a difficult time will think, "I am happy that I live here."' This positive attitude was not unique to Mr Kamiyama; he said that he felt that it was the general attitude of people in the summer of 1999. Those working for local community activities in PRH said several times that they were no longer in a post-disaster emergency situation; they had returned to normality and they had to stand on their own feet. However, there were a few who complained and asked for more government subsidies.

Mrs Sakura was often critical of the media for the tone of its coverage and the questions that reporters asked those in the community, especially when they focused on elderly and unhappy people. She criticized them for reporting bias and ignoring the true situation. On 4 February 1999, she said,

> The size of the housing this side is larger. This is for a family. Because these were planned before the earthquake, when this was completed, the city allocated this to a family first. I know the mass media complains that many older people were left in temporary shelters. But I believe it was good to move a family first. I see things from inside. A family cannot live in a temporary shelter room for that long. It is built for two years. It is limited. It may be hard but it was reasonable to ask an older person living alone to put up with the single room temporary shelter for a little longer than a family.
>
> In a temporary shelter, a family of four members lived in one shelter. Facing each other for a long time caused problems such as domestic violence. They needed to help to get them out first. I think it was right that the public administration move them first into the large flats which were built first. The mass media say that it is was cruel for the public administration to leave those

> old people living alone in a temporary shelter that long. I do not agree with the media coverage.

Annoyed by the media coverage, she added on 16 February 1999,

> NHK (a station similar to Britain's BBC) comes to cover here as a one-time shot. It is so annoying. They come and ask, 'Is there anyone who is miserable? Depressed?' They already have their own scenario in their mind. They should see the reality and facts. They should take a long time to do that. They try to find things by their *atama* (brain, mind), not by their *ashi* (foot). We are the ones who they should feel sorry for.
>
> Three years ago, I was sorrier for the family of four or five living in a temporary shelter than those old people living alone there. When I visited around, there were secondary school pupils throwing stuff at their mother. The children were not bad children, but it is so irritating to live as a family in such a small shelter, facing each other for such a long time. No space. I fully agree that we provide housing for them first. I do not agree with the mass media coverage, 'Oh, poor old people left alone in a temporary shelter'.
>
> Senri New Town in Osaka must be the same in terms of ageing, but it will have gradual change and then they have some base to cope with the change. But here this aged society is brought in, as one, from the beginning.
>
> The other day, a professor was brought here by NHK to deliver a lecture. He came by car and went by car. His talk was OK but people here were saying, 'They do not understand. It is something beyond our reach. Take a long time to know the reality and find out the reality and write a good report'.

What is shown on TV does not necessarily reflect the reporter's intentions. In TV station planning committees, each reporter has to fight to put a report onto the agenda. In an interview on 31 May 1999, a reporter at NHK Kobe said,

> What is happening here is experimental. It is worth getting attention from all over Japan. But the issues related to the Hanshin earthquake are dealt with as local news. It used to win the attention of the central station of Kobe but no longer. It used to be given priority, whatever we wrote and reported, but now no longer. I am unhappy with this attitude of the central station. I do not understand why they treat our news as local news. This should be national news. This is what the whole of Japan has to look and think about together towards the ageing society.

Among all the information they collected, what was taken up on air was decided by the TV station committee. Furthermore, a local TV station not only had more coverage of the earthquake affected people, but the tone of the messages were often more sympathetic to the people affected by the earthquake and more likely to blame *gyōsei* for problems.

### *Measures of success*

**Former elderly residents of TSH**

This section discusses the changing media focus over time and the way that this often changed the perceptions of elderly people after moving from TSH to PRH. Former elderly residents reported missing several things they had enjoyed at TSH, but these statements changed over time.

Elderly residents had grown accustomed to TSH and had, to some extent at least, made it into a home. The anxiety of moving to a new place they did not know often influenced their perceptions of PRH. Furthermore, PRH residents did not receive the special treatment and post-earthquake relief as they did in TSH. They were also faced with higher costs at the same time as TSH was free and PRH only subsidized.

### *Gyōsei*

The public administration's measure of success has shifted significantly from quantity to quality, that is, the amount of housing to the quality of life that the residents could expect from each type of housing. In the first year following the earthquake, the most urgent task of the public administration was to provide TSH to those who had lost their homes. To achieve a quantitative measure was a priority. As time went on, more attention was given to the quality of life in these homes, partly in response to public opinion, criticism from the media and workers and the findings of the survey they conducted on the people living in the homes.

As the measure shifted to the quality of life, the public administration started a new scheme, helping people to use both the existing health and welfare system and the new schemes funded by the special post-earthquake reconstruction fund. They planned and built Silver Housing, public housing specially designed for old people. They also increased the number of LSAs and hired new staff to visit elderly people daily and help them reconstruct their lives.

Kobe was one of the first cities to build capacity through local government initiatives (Kose, 1998: 125–139). The earthquake resulted in new bottom-up approaches, rather than top-down approaches from the central government being adopted.

Life care communities were constructed by several local governments. Kobe City built one of the first of these. The project contained a special nursing home and a day care center on the lower floors, with 31 units for independent elderly persons on the upper floors. One unit was provided for a LSA.

The floor area of a single person's unit was $35m^2$, with two rooms and a kitchen. The unit for a couple was $48m^2$, with two rooms and a dining-kitchen. The LSA would periodically check the safety and health of the elderly resident and provide help in case of need. The LSA also acted as a consultant for daily living to improve the resident's quality of life. The day care center provides services such as meal preparation, bathing, rehabilitation and recreation. The nursing home could also accommodate short stays in case of need.

**PRH community leaders**

Mr Kamiyama agreed with the media that new PRH residents felt depressed and isolated by the modern environment when compared to the crowded TSH. He said,

> First they were happy to move into permanent housing. Then they realized that they were shut in by the heavy iron door. The temporary shelter was not well-built (*gata gata*) but there was a closeness to other people. In a good sense, this is well-built (*gacchiri*). Yet the sense of isolation and loneliness increases here.
>
> When the strong winter wind blows in the direction opposite to the door, old people cannot open the heavy iron door. But, they also cannot sleep at night because they remember the sound of closing the heavy iron door (*gan*!) and they complain.

Although things had improved over time, he found that building community presented other challenges, including how strangers can be encouraged to help each other and to be independent at the same time.

Various actors like Mr Kamiyama, Mrs Sakura and Professor Konshi said that it was difficult to develop community in such a mammoth housing complex. Community development does not work well if the community is too big or too small. If it is too big, it is difficult for community members to reach an agreement and those who do not participate can be excluded. But if it is too small, it might not have the dynamism to maintain the momentum of community activities.

Also, when the size of the community is too big, it is difficult to form and manage a self-governing body. If a self-governing body is not formed, the community would not have an official representative and it would be difficult to collaborate with the public administration and other representatives, a critical impediment to community development.

More agencies are involved in larger communities. My fieldwork identified numerous actors: the *minsei iin* (community welfare commissioner), the *jichikai* (self-governing body), the *gyōsei* (public administration), the *jūmin* (residents), volunteers, women's groups, NPOs, and the media.

One of the features of PRH was its large size. To meet the enormous demand for apartments after the earthquake, a group of high-rise buildings with ten to eighteen floors were built in Kobe City and other cities in the Hyogo Prefecture.

Mr Takada was a welfare commissioner at Natsu Aki after being a welfare commissioner in the Tarumi Ward in central Kobe City for twelve years. He said,

> When I moved to the east side of the new town of the Natsu Aki before the earthquake, I thought I was going to retire and spend time by myself at the nice new house I had bought. But Kobe City contacted me to ask me to work as a welfare commissioner again, as they thought it would be better to have a male welfare commissioner and I am experienced. But the work is much harder than the previous place. In the Tarumi Ward of Kobe City, there were not many older people living alone or bedridden. In other areas, these kinds of people may increase gradually. But here, it was like one day suddenly this large number of households showed up.

Not all the residents at the field sites I visited were able to participate in the activities. The community center was large enough to contain one-tenth of the residents. In response to this, Mrs Sakura did not try to cover everything at once but allowed activities to evolve gradually over time.

## Volunteers

Volunteers played a major and dynamic role in TSH community development.

After the earthquake in Kobe, people discussed the Western notion of community. There was an NGO named the 'Community Support Network' and several symposia on community development were held in Kobe.

Studies suggest that there is a relationship between social support and physical and psychological health status (Bowling, 1991a: 68–83). In the Kobe case, support networks can help to replace lost friends and family members at the same time as creating new homes and communities. In

Kobe, where Christians are not a prevalent group, examples were taken from the Western context; study tours often focused on hospitals where medical staff and volunteers from a Christian church worked together. In Japan, Buddhism and Shintoism are more common than Christianity and their temples provided such functions. However, such practices no longer serve the community much in modern urban Japanese society in the way they do in some other Asian contexts such as Thailand. There are studies and reports in recent years on how Thai Buddhist temples are taking major roles in providing community care and increasing awareness for HIV and AIDS (UNICEF, 2001).

Most of the volunteer work was done at TSH, giving residents a feeling of being cared for. Human contact was most frequently reported as missing by PRH residents. For various reasons, volunteers did not work in PRH to the same extent that they did in TSH. It was not thought that PRH needed emergency relief after the earthquake, that life had returned to normal and residents were independent. After five years, government subsidies for volunteer activities ended and further extensions were made year by year.

It was also easier for a volunteer or neighbor to make a visit in tenement type TSH housing rather than in high-rise modern apartment PRH buildings, with heavy iron doors (*tetsu no tobira*). Once the door was closed, the resident was shut out from the outside world and could not hear their neighbors. Through the windows, PRH residents can see only sky, whereas they were used to see people passing by in TSH.

Although there were fewer volunteers in PRH, their activities were replaced by LSAs, who received a salary. However, LSAs did not always work as well as the volunteers did, despite their more official status. This might be because the number of LSAs was much smaller than the number of volunteers. The extent to which a LSA can reach out and give cover is limited. It might also be because some LSAs were too young and inexperienced to develop trusting relationships with the elderly residents. It is important to share experiences when building friendships in a social work context (Bornat, 1997: 31).

Although the role of LSA was official and received a salary and an office at the PRH, it was more stringently regulated than the volunteers at TSH. These regulations could present a barrier to developing human relationships at more than an official level if this was what the residents wanted. Many staff said that they had to learn how to manage their relationships with residents because they were not themselves residents and would leave when their contracts ended. Emotional attachments would make leaving more difficult for both staff and residents.

Regulations prohibited LSAs from receiving gifts from residents, even a cup of tea or a sweet, so the residents would not compete with each other. One day an old man gave a small gift to a staff member, saying that his only daughter died in the earthquake and he had no family and no one to buy a souvenir for after his trip to a hot spring. The staff member could not refuse the gift, illustrating how the regulations were sometimes difficult to meet.

## Community and public health

### *Health promotion*

It is a common assumption of officials that researchers are mostly interested in the area of health promotion and community development. Health promotion in PRH compounds was carried out in collaboration with the Department of Health of the ward of the public administration.

The Japanese are highly health conscious. They enjoy the longest average life expectancy in the world. Health is a major part of self-sufficient and independent life. Elderly people often see health as the main determinant of the quality of their lives and many make arrangements in preparation for the event of something going wrong (Wilson, 2000: 15). Health promotion is often an effective pathway for community generation. In my research, officials and community leaders, such as the *jichikaicho* (chief of the self-governing body), would often worry about elderly people living alone in their communities and health promotion made it easier to get social support to launch and implement activities. They could collaborate with the Department of Health to enhance health and welfare work. Such collaboration also often included *gyōsei* if they had the necessary accountability, transparency and capacity, as well as a *nin-I dantai* (voluntary group) or *jichikai* (self-governing body). This put the responsibility on the local community to prepare for intervention from government.

Health promotion programs have other benefits in addition to enhancing the health of residents. They can facilitate community development. Community development and health promotion programs have a strong correlation.

### *Nutrition*

In the Hyogo Public Health Survey, special attention was paid to the diet and eating habits of elderly people living alone. Education about good eating habits receives a lot of attention in Japan in general. A good diet is considered to be difficult to maintain when an elderly person lives alone, especially in a suburban area away from markets and restaurants. In the survey of both TSH and general housing, the proportion of illness resulting from poor diet was higher among elderly people, especially those living alone.

In an interview on 26 January 1999, the Director of the Department of Health of Kobe City explained,

> We provide a place and an opportunity for community development through a diet class and a physical exercise class. We invite people from the local community to work together. Among those who are doing fine now, as they tend to be old already, we are concerned with any change in their condition which would happen in the near future. We believe that community development is a must. The community needs to develop a base and a capacity to accept those who can no longer be independent, those who have lost their health and have a difficult time interacting with their neighbors. Now we can develop a system to help them. We incorporate *isobi* (play, games) into our health promotion program, such as *origami* (paper folding). We cooperate with the *jichi-kai* (self-governing body).

When someone lives alone, they tend not to eat regularly or have a balanced diet. In Japanese society, eating well is considered to be elementary for a healthy life. There is a Chinese-Japanese saying, *ishoku dōgen* (food and medicine from the same source).

Older people tend to eat less than younger people (Anastassiadou and Heath, 2002, Saava and Kisper-Hint, 2002). Medically, daily nutrient intakes are lower in those who have fewer natural or functional teeth and ill-fitting mandibular dentures than those who have more teeth and have no such problems (Marshall et al., 2002). Older people might also eat with fewer people present (de Castro, 2002). In my fieldwork, some elderly people mentioned that when living alone or as a couple, they tended to cook one dish at a time and eat the same dish every day for a while; they tend to consume a more monotonous diet (Fanelli and Stevenhagen, 1985). Even in prosperous countries where malnutrition has supposedly been abolished, it is possible to be undernourished in later life (Wilson, 2000: 105). Although elderly women often perceive eating in fellowship as a pleasure, they tend to simplify cooking and eating when living alone (Gustafsson and Sidenvall, 2002). The same was mentioned by older women living alone at my fieldwork sites. Physiologically it is also known that weak appetite in elderly people is tied to a hormone known to suppress appetite (Science News, 2001, MacIntosh and Morley, 2001, Amella, 2002).

Japan has four seasons and each season has special foods. The Japanese enjoy being gourmands, so programs that promote healthy eating attract people's interest. The basics of nutrition are taught in primary school, but poor elderly people might not have completed primary school due to the war. Nonetheless, there are many cookery programs on TV and they are always popular. Cooking and diet articles also frequently appear in newspapers and magazines. Such emphasis on a balanced diet could be an important factor in why Japanese live so long (Cockerham and Yamori, 2001, Suzuki et al., 2001).

Concerned about the health of elderly people living alone, the Department of Health in collaboration with the self-governing body (or another representative if no body had been formed) organized monthly health promotion programs at PRH communities. The purpose of the programs was to increase the awareness of the importance of a daily balanced diet to maintain good health, and to encourage elderly people living alone to follow a healthy diet. These activities also provided an opportunity for residents to come out from their apartments.

When talking about the health promotion activities at Fuyuyama PRH, the chief of public health nurses at the Department of Health of the Central Ward of Kobe City said,

> This program is organized by the Department of Health. The Department of Health does *junkai sōdan* (patrol and counseling). This is a kind of *fureai kenkō sōdan* (human contacts and health counseling). A doctor also joins in. Today's program is for old people living alone. We show how to eat a healthy diet, with real dishes as examples. We cover counseling in the areas of health, medicine, *kokoro no care* (mental care), and nutrition. One small dish is counted as one point. At the reception desk, an old person is given a sheet of paper with the chart of a nutritious diet and a table displaying each kind of dish prepared in that day's program. Afterwards, we do *kenkō taisō* (heathy physical activity).

She added,

> These are health promotion programs, but not only for health. This is a place for community development.

Such programs were aimed only at households of residents aged over 65 and living alone. Those who lived and ate alone were particularly likely to lose interest in cooking (Wilson, 2000: 105). People who live alone tended to eat the same food and often have an unbalanced diet.

The chief public health nurse of the Department of Health of the Central District of Kobe City said at a health promotion program at the Fuyuyama PRH, 25 February 1999,

> We can do these activities thanks to Mr Tanaka, the chief of the self-governing body. Without a representative from the community, we cannot just go in and do this. He makes things much easier for us to carry out these programs. He did foundation building before we came in. We cannot do everything because our resources are limited.

## Elderly people living alone

*Hitorigurashi* (living alone), or *dokkyo* (solo living), is an important factor to keep in mind when evaluating housing impacts (Iliffe, 1992, Rubinstein et al., 1992). This is especially so when looking at housing for elderly people in the Japanese context. Japan has an exceptionally high cohabitation rate of elderly people with their children when compared to other industrialized countries. Yet population ageing and various social changes have coincided with a decline in cohabitation. More elderly people than ever will be living away from their children in the coming years, putting greater reliance on government support (Campbell and Ikegami, 2000). In 1970, elderly Japanese needing special attention from expansions to old-age policy were commonly referred to as *hitorigurashi* (living alone), a category that in common parlance was blurred with elderly couples living independently (Campbell and Ikegami, 2000: 86).

My research found that the categories for targeted groups of weak and vulnerable people remain the same across surveys of health and welfare. Surveys like the Hyogo Health Survey, two reports from the Department of Health of the West ward of Kobe City and media data all included categories for elderly people and elderly people living alone.

### *Loneliness*

It seems that residents in poor housing conditions often go out and socialize, while those in good conditions stay at home and become lonely. The media often emphasized 'loneliness' to explain why life at PRH was worse than life at TSH. Some people see 'old age' as a factor in 'loneliness'. Chapter seven will look at these issues in more depth.

Life-style differences between old and young people might explain different responses to housing styles. Younger people who go out to work during the day might feel more comfortable living in a high-rise apartment building in an urban city. However, things would be different for older people who spend most of their time at home. Community life might be more important to them.

Insecurity and loneliness may be the main push factors that make elderly people willing to move into some form of institution (Wilson, 2000: 151). This grouping of isolated elders into collective houses was also observed in Kobe. Some made friends with other residents and were able to lessen their sense of loneliness, although elderly friends often became unwell and moved into a nursing home. For some resident, collective housing offered a sense of security rather than friendship.

### *Working together*

It is often argued that it is necessary to involve young and healthy people to support activities for elderly people. Often, elderly people regret that the younger generation does not respect elderly people in the way that they did when they were young. But the Kobe case has shown that older people need not necessarily be taken care of by young and healthy people. Several people, like Ms Kuwahara, a LSA in a group home type TSH, Mrs Nakayama, the director of the nursing home in Ashiya City, and various TV programs explained that elderly people who are still able to help others can take on this role instead. Elderly people can live happily and with a little assistance they can stand on their own feet. This way they feel they are contributing to the community and feel life is worth living. Mrs Sakura seemed to believe that anyone, no matter how old or disabled, should be able to take an active role in community care development. She encouraged elderly residents and disabled people to participate in community activities and in the self-governing body.

### *Housing*

Home ownership is often the determinant of wealth and quality of life in old age. (Hughes, 1995). Home ownership provides protection and opportunities.

Special housing and communities for elderly people are a recent phenomenon and a major concern for ageing societies. Various attempts have been made and it is expected that there will be more.

Various studies from the United States and Europe are available on this topic (Drake, 1998, Steinfeld, 1981). Highly visible segregated housing for the elderly became a significant component of the urban landscape in the United States after the Second World War (Rowles, 1994). As a result segregation increased:

> While the majority of older people continued to live in community settings, the proliferation of alternatives, particularly high-rise edifices in the central areas of large cities, began to convey an image of the elderly as somehow separate and in need of special services and housing arrangements. Although the actual number of older people involved was small, the high visibility of the elderly high-rises conveyed an image that the normative pattern for the elderly was the abandonment of their homes and a movement into 'supportive' settings where they could obtain types of assistance that were not available in their homes. Reflecting the ethos of the time, many of the housing and long-term-care options for the elderly assumed a willingness to relocate (Rowles, 1994: 115–126).

Steinfeld (1981) analyzed the meanings of housing in old age. He discussed housing from several aspects, including housing as a symbol of self in old age, housing as a symbol of social change and housing as a shaper of identity. Most housing is built as a type of dwelling, for example apartments, town houses, single-family detached dwellings. Housing for the elderly, however, is built as a type of dwelling (usually apartments) that is exclusively for a particular group of people.

Housing for the elderly is an important determinant of access to health care. Old people's homes may provide some functions similar to a hospital On the other hand, the selection of housing is not determined only by the convenience and security of access to medical care, as Palmore notes:

> Because housing is more than a place to live, it can have many functions: a symbol of independence, a focal point for family gatherings; a source of pleasant memories; a link to the community; a focus of useful activities such as hobbies and crafts, gardening, and home improvement. Because of these many functions, age segregation in housing can have major consequences (Palmore, 1999: 140–151).

Health in later life can be greatly improved through housing policy (Wilson, 2000: 106).

## Summary and conclusion

Because of the need to provide 40,000 PRH apartments in a short time, there was little choice for the *gyōsei* (public administration) but to build high-rise buildings. Their measure of success was the amount of housing they were able to provide in the early stages of the reconstruction process after the earthquake. The impact of the housing type on the quality of life started

Photo 6-3: Community center in the Natsu Aki PRH complex

to receive attention from *gyōsei* in the later stages of the reconstruction process. This was partly due to the influence of the media; public officials were sensitive to what was reported by the media. It was also partly due to communication between *gyōsei* and the local community. The change in perception was led bottom-up by staff, like public health nurses, and others working at the local level, like the self-governing body and *minsei iin* (community welfare commissioner). It can also be said that the *gyōsei* was sensitive to the needs of communities.

When one refers to housing as TSH or PRH, they do not mean just the housing type itself, but the whole community type. People prefer access to services, resources and opportunities. When community development cannot be established, no matter how nice the apartment provided, the housing experience is not good. The leader of the volunteer nurses working at a TSH said, 'I do not understand why an old person living alone moves to public reconstruction housing. What are they going to do there?'

PRH are high-rise modern apartment buildings with a high proportion of elderly people living alone in an urban environment; they are a future model of an aged and urbanized society. Society can learn from the experience of community generation at the TSH in order to incorporate, with adjustments, what worked in low rise type housing into the new experience at the PRH.

This chapter drew attention to six issues relating to the media and PRH. First, conflicts were observed between local media and the central

Photo 6-4: Older residents playing ground golf at the park next to the Natsu Aki PRH complex.

government, public and private TV stations and national and local TV stations over what to put on the media agenda.

Second, although the media did not talk about the subsidies supporting PRH residents, rent and personal finance were major issues for people in my fieldwork. Many at PRH were happy about having a nice apartment that they would have never have been able to afford if not for the post-earthquake public housing schemes, even though there was a huge gap between the number of destroyed houses and the number of PRH provided. Those who did not move to PRH were dispersed into non-subsidized general housing, some by their own choice when they gave up on winning the PRH lottery. Priority for PRH accommodation was given to those who would otherwise be unable to afford to move on their own.

Third, the media's perspective on things did not always reflect the perspectives of people I observed in my fieldwork. Residents and staff of PRH were more positive about life in PRH while the media were more critical. The media's tone changed in 2000, partly because it was the fifth anniversary of the earthquake, a time of *kugiri* (separation time) and *fushime* (bamboo joint). In Japanese culture, the fifth anniversary is often seen as a time to start a new outlook.

Fourth, media reports did not necessarily interpret the views of residents and staff in the right way. The media interpreted the anxiety of residents about life at PRH as criticism of the public administration, housing design and so on. It did not take into account that residents were going through a transition from emergency relief to normal life.

Fifth, although the media focus was on elderly residents of PRH, it was not intended as an old people's home. In consideration of the proportion of elderly people living there, the administration built Silver Housing, but over time as the elderly people die younger people might move in. The apartments for single people were designed to be easily renovated to combine two or three to make a family apartment. In this event, PRH might not remain an example of Japan's ageing society.

Sixth and finally, while the ageing population is an issue of concern, elderly people may not be the most vulnerable group. Those in their fifties may be more vulnerable as they are not old enough to receive the benefits that elderly people receive and it was often difficult for them to find a new job if they lost their workplace or business in the earthquake. In Japan, job advertisements often stipulate that applicants should be younger thanr 35.

Community development is not adversely affected by the age of residents, but communities with more elderly people may find it more difficult to generate a *jichikai*. However, there are other factors that affect the extent of a resident's participation in community. Other factors include their previous experience at TSH and the size of the community, as well as a number of community characteristics like leadership, communication, collaboration, resources, and the extent of interpersonal networking.

The *renkei* (coordinated collaboration) between the *gyōsei* (public administration) and local communities was the key to developing community care. Often the *minsei iin* (the welfare commissioner, official volunteer) had already established a trusting relationship with residents. The *jichikai* was also important. The *gyōsei* cannot work in a community that is not active or has no representative body with leadership and coordination.

It was more difficult to generate community in large compounds. Many staff felt that people behaved irresponsibly in large groups and consensus was difficult to establish.

In many cases, health promotion programs initiated by the Department of Health served as a catalyst for community development. Health is considered to be a major part of self-sufficient and independent life in Japan and it is easier to get social support to launch and implement activities in the health field.

The role of government in supporting elderly people is controversial. Local, municipal, and central governments have a major role to play in assuring a healthy life for their citizens. During the earthquake reconstruction process, Kobe City became the first local government to cooperate with local communities to produce life care communities. PRH community development is one of these initiatives. Although the government cannot provide enough Silver or collective housing, Hyogo Prefecture has far more public housing than any other prefectures since the earthquake. Many of these are very modern apartment buildings.

In Japan, volunteer activities have often been seen as requiring self sacrifice. The earthquake changed this view, making Japanese people think of how to contribute to society to make life worth living. When the TSH compounds were closed, volunteer groups who could still offer their time looked for a PRH where they could continue to volunteer. This sometimes created tension between different volunteer groups. Also, there was some resistance from residents and staff of PRH to outsiders.

According to the US Centers for Disease Control and Prevention (CDC, 1997), processes for community engagement are more likely to be successful when true equality of community leadership is intended and applied, as opposed to an authoritative or top-down approach. From the experience in Kobe, it has not yet been figured out what approach will work in Japanese society.

It also states that the community creation process can be difficult and labor intensive and requires dedicated resources to help ensure success, including time, money and people. On the basis of numbers and the size of the compounds, the two PRH leaders discussed were not all that successful in community creation but it was surely much better than nothing. The two PRH compounds that I conducted my fieldwork in were the most active ones among all other 76 compounds. More needs to be done in the area of community development at PRH.

# Chapter Seven
# Loneliness

'I am lonely' was the most frequently used expression of elderly residents of TSH and PRH. This chapter explores what it means.

In the first few years after the earthquake, the success of the recovery was measured by the number of houses provided by the *gyōsei* (public administration) to those who lost their homes in the earthquake. But as time passed, the measure of success shifted from hard, quantitative facts to soft, qualitative ones. The media often emphasized that PRH residents felt lonelier than those at TSH. A news item broadcast on ABC TV, 13 January 1999, said that Mr Keisuke Kawabata (60) had felt happier in TSH than in PRH:

> When I lived in a temporary shelter, someone always dropped by and offered to share some food: 'I made miso soup. I made *nikujaga* (Japanese traditional home cooking with beef and potato).' But now here in public reconstruction housing, I often feel lonely. There are still days when I have no exchange of words with anyone. When a person gets older, it is not easy to make new friends like in their younger days.

Stevens (2001: 183–202) wrote, 'Friendship is a type of relationship that can become difficult to sustain as people age. Age-related changes in health, relocation and retirement influence older persons' access to actual and potential friends'.

Residents often appear on TV saying things like 'I am lonely' 'I am alone', 'I have no one to talk to', 'It is difficult to make friends here', 'I have no husband or wife'. In this sense, loneliness conveys the feeling of not having company.

## Loneliness and elderly people

The media reports often related 'loneliness' to 'old people'. In his comment above, Mr Kawabata explained how he sees his old age as a factor in his struggle to make new friends.

The tone of the media in January 1999 implied that residents were lonelier in PRH. The reports tended to blame *gyōsei* for planning and providing inappropriate housing designs, especially high-rise buildings. The media blamed the loneliness of elderly people who lived alone on the perceived failings of the post-earthquake *seikatsu fukkō* (life reconstruction).[1] It even claimed that some elderly people wanted to return to TSH, even though the TSH had been pulled down. It said that others wanted to return to their original homes that had been destroyed in the earthquake. However, during my fieldwork I observed that the desire to leave PRH was often derived from feelings of anxiety and uncertainty that occur at a time of the transition of starting a new life in a different home. Although many felt lonely remembering the good old days, they accepted that that time no longer existed. To explain this feeling, Mr Tanaka of the Fuyuyama PRH would quote from an old song called '*Furusato* (home town)': '*Usagi oishi ano yama* (Running after rabbits in that mountain)'.

Most elderly residents admitted loneliness but accepted it as a natural process. It did not mean that they were not looking forward to life in PRH. Indeed, these feelings were later reflected by the media, which became more positive in January 2000.

In the literature on elderly people, loneliness is an importance theme. The pathology model of ageing focuses on poverty, bereavement, social isolation, loneliness, role loss, illness, handicap, apathy and abuse (Onyx and Benton, 1995, Fenneal et al., 1988). Loneliness is a serious problem in old age, whether married or not, and is an important issue to look at in relation to providing housing for the old (Heywood, Oldman and Means, 2002, Gibson, 2000, Holmen, 2002, Stevens and Tillburg, 2000).

Although loneliness is a big problem for those who have lost someone who shared their home, adults who are more used to living alone can over time suffer worse loneliness as their ability to go out or to invite people diminishes, or the loss of hearing or sight reduces their ability to be sociable.

## Loneliness and social isolation

Enomoto (1996)[2] writes,

> The method of link services has benefited Icelandic elderly persons greatly, improving their security and safety, increasing their opportunities to exchange with others and enjoy themselves, relieving them from feeling isolated, lonely, or useless, and encouraging them to lead self-supporting lives, feeling that life is worth living.

'Link services' were developed at an old people's home in the Iceland. The word 'link' here is similar to the Japanese word '*tsunagari*' (human relationships). Enomoto's passage contains several key words and phrases that can be found in other literature and in my own research in Kobe. These words and phrases include, 'an opportunity to exchange with others', 'isolation', 'being useful to others and feeling one is still useful' and that life is 'worth living'.

Giddens argued that,

> Personal meaninglessness—the feeling that life has nothing worthwhile to offer—becomes a fundamental psychic problem in circumstances of late modernity (1991: 9).

Hochschild (1973)[3] studied an 'unexpected community' of 43 old people who solved the problem of their loneliness by living together in collective housing. She illustrated cases where isolation led to loneliness. Themes of separation and integration also came into her discussion.

In my study, elderly residents experienced discrimination and segregation. These experiences could breed greater feelings of loneliness. An old woman at the Fuyuyama PRH said, 'Living here, the people around us say, "They are paying low rent". No neighbors from the original community talk to people like us. I feel so miserable. I often feel I want to die soon'.

Hochschild (1973) suggests that there are problems that specifically affect elderly people, such as poverty and poor health. The media coverage and the *gyōsei* (the public administration) who I interviewed thought that elderly residents of TSH and PRH had similar problems. While this may be true in general, there are people in their fifties and sixties who are less healthy than people in their seventies. Some say this phenomenon is caused by the survival effect (Bowling and Brown, 1991, Bowling et al., 1997).

A 60-year-old man who I visited at the Haruyama TSH was alcoholic and died before I finished my fieldwork. He had been a cook at a restaurant and lost his workplace in the earthquake. When I visited him in April 1998, three years later, he was drinking but not eating at all. His room smelled bad and his bathroom was a mess. Because he was a war orphan, he had lived a difficult life. His parents had died in the bombing towards the end of the Second World War. With the help of a volunteer nurse, we cleaned the bathroom and the main room. We cooked for him but he did not want to eat alone. He wanted us to stay while he ate but we had to go and visit others and as volunteers there were limitations on what we could do to help. He was losing the will to live but he was still interested in reading a cookery book. He died of malnutrition related to alcoholism, despite living in a relatively affluent society like Japan.

Eating alone can be seen as a symbolic manifestation of loneliness. According to Hochschild (1973), elderly residents in his study often spoke of having meals with neighbors, although this didn't include help with household chores like borrowing a butter knife or asking for a telephone number. The difference between eating alone and eating with someone kept coming up in my fieldwork.

Hochschild (1973) wrote that while health and housing are basic problems, more serious than this is the feeling of rejection a person experiences if they think they are not needed by society. In my fieldwork, elderly people were happy if they felt proud of their children. If they didn't, or if their children didn't visit them, they often looked and sounded desperately lonely. Some of them blamed their daughters-in-law and sons-in-law if they had relationship problems with their children.

In some cases, people started to have serious relationship problems with their family members after the earthquake. Some discovered they had different ways of coping with the emergency situation and often priorities changed. Some lost trust in each other because they were not rescued quickly from their destroyed houses, or because their family members had not asked for them immediately after the disaster. These experiences can cause people to close in on themselves. In my fieldwork, I heard both stories of family breakdown and the reaffirmation of family ties. Some were divorced after the earthquake, others got back together to support each other through the difficult times. Some people met and got married while working together to rebuild communities and to help those affected by the earthquake.

While loneliness often manifests when a person is alone, just because a person is alone doesn't mean they are lonely. Likewise, a person who is surrounded by people, even their own children and grandchildren, can be very lonely. Perhaps no one understands them well, even though they enjoy good relationships with others.

Hochschild's (1973) results differed depending on her methodology. In her surveys, half of the respondents answered that they were alone. During her field visit, she observed many of these elderly people receiving phone calls, leading her to ask if it could still be said that they were alone.

In the Japanese media, elderly people were seen as *jakusha* (the weak). They were seen as people who required special care, were lonely and unhappy. This was particularly evident in the titles of TV programs. For example, a NHK special feature that aired on 14 January 19999 was titled, 'Isolated old people: Report from a public reconstruction housing project after the earthquake'.

In an interview on the 26 January 1999, the Director of the Department of Health and Welfare of Kobe City said,

> For the post-earthquake public reconstruction housings, priority is given to the weak (*jakusha*) such as old people, disabled people, and those on low income with many children.

This was acceptable to the public and resolved any criticism of *gyōsei*. But after the main part of his talk, when it became more informal, he said frankly,

> Employment has dropped significantly in Kobe. We need to promote employment. This is what I believe we need to give priority to. We need to create jobs for those unemployed in Kobe, to spend money on small things. I believe we need to spend to create jobs. It is the time for job creation. As a post-earthquake reconstruction issue, job creation is now a priority. You know 340,000 households lost their homes (28,000 apartments were built). The rest are trying to stand on their own. They have large debts. They are paying a monthly rent of JPY150–180,000. Young couples with small children also need money.

Here the director was proposing shifting the focus of public support from elderly people to support for the middle generation. His comments anticipate changes in the tone and focus of media coverage from 1999 to 2000, almost a year later. The shifts were evident in the interactions among (1) the tone and focus of the media, (2) the voice of *gyōsei* regarding the priorities of policy direction, and (3) the voices heard in my field work.

Generally speaking, those who moved into PRH were working class. The chief public health nurse said in response to the director's comments above,

> They have modern equipment [in a PRH apartment]. It is a modern bath [rather than traditional Japanese style]. It is a Western bath with Japanese modern technology. You can fix a bath with computerized buttons. They cannot figure out how to use them, so a support staff or advisor needs to explain it to them. With this type of bath, you do not say, 'Heat the *wakasu* (the traditional Japanese style bath)'. But you say, '*Yu wo haru* (fill up the bath tub)'. You know...those who did not have their own private bathroom at home and went to a public bath before the earthquake.

This reveals some of the daily experience of working class Japanese. But in the meeting, they did not call them the poor. As 90% of Japanese people consider themselves middle-class, it might not be culturally appropriate to openly distinguish one from the other by economic status. The local government officials avoided calling these people poor in the economic sense, but many are defined as poor because of their age. In my fieldwork, there were people who seemed to have been wealthier when they were younger. This was a point that TV programs also often raised.

One of the Hyogo Prefecture Health Survey questions asked respondents about the causes of stress and how they managed then. 25% of TSH residents and 15% of PRH residents selected a multiple choice option that said they felt that they spent a lot of time alone. Those who were of poor health were more likely to select this answer, whereas those who felt they had good health said that they worked hard and enjoyed leisure activities.

## Loneliness and lack of meaningful relationships

For elderly people who do not have work or family obligations, their place of residence and neighborhood friendships are more important (Hochschild, 1973). Those elderly residents housed alongside young people experience more isolation due to the mutual lack of interests.

The place of residence is more important to the lower class than the middle class (Hochschild, 1973). I observed a similar situation during my fieldwork. The loneliness among lower class residents of TSH and PRH was clear. In comparison, in the more affluent areas of Nishinomiya and Ashiya, detached individual houses were rebuilt within a year. Hirayama (2000) called this phenomenon the 'polarisation of housing recovery'. He showed that the peak of building activity was July 1997 and then it slowly started to decline (Hirayama, 2000). Town scenery was significantly different in the more affluent areas than in the areas where TSH and then later PRH were built. This differentiated between those who had money to rebuild their house and those who did not. For example, middle class people were not used to living in single floor *Nagaya,* shared by several households. They had their own circles of friends with similar backgrounds and interests. They did not need to live in a neighborhood to form a community.

The first two years of the Hyogo Health Survey drew respondents from general housing as well as TSH and PRH (first featured in the second year). In the third year the survey was not conducted in general housing, reflecting the view of the Department of Health that middle class people not living in PRH no longer needed public aid, but had recovered and returned to normal life.

## Being alone and feeling lonely

If an elderly person lives alone, it does not necessarily mean that they are cut off from friends, neighbors and family, especially children. Hochschild

(1973) suggests that most elderly people do not want to live with their children. Hashimoto (1996) demonstrated that Japanese elderly people prefer to live separately from their children so that they gain independence and are not a burden on them.

Hashimoto (1996: 8) presented the case of a Japanese woman in her late 60s who was still professionally active as a nurse. She said, 'People are always asking me if I don't get *sabishii* (lonely) living alone'. In the Japanese value system, it seems to be common to put living alone and loneliness together.

The purpose of the lunch service in the housing compounds was to make sure that residents could have a communal meal as part of a balanced diet. The media treated a person's act of eating alone as a sign of them having no friends or family. Eating in one's room was seen as unhealthy and some would say that eating without exchanging words or sharing food would make their food tasteless.

People feel lonely without the opportunity to offer hospitality. During my fieldwork in TSH, I heard about elderly people who enjoyed cooking food to share with their neighbors. The ability to grow vegetables (although this was illegal) and pick mountain food to share was one of the aspects that made many residents happier to live in remote and mountainous TSH than in their downtown homes.

A few TSH homes were featured by the media as examples of successful cases. But a group home with no large rooms for common activities was not comfortable for permanent residents. Common rooms would enable residents to exchange *fureai* (human contact) and *tsunagari* (human connections), and is therefore important for community generation. One program featured a private group home for eight elderly women built after the earthquake. The elderly women were happy because they ate with each other every day.

## The sense of loneliness

Residents of TSH and PRH have in common their loss of housing due to the earthquake. Many also lost their work and lifestyle. Some even lost family members.

Sun-TV followed Ms Chieko Ueda, about 70, from 1995 to 2000 for a program that was broadcast 7 January 2000. In December 1999, she said,

> I have lived in a temporary shelter alone for three years and five months. People there have experienced the same kind of hurt feelings; they can understand. They care about each other. They were gentle.

> door, I sit alone in my home. I want to live with my daughter. You know...I live in building Number 2.

This is a case of an old woman who had lost her daughter to her son-in-law. They once tried to live together but it did not work out. By saying that it was her who initiated the move, Ms Sato was trying to say she had not been rejected by her daughter and her family. Despite living in a nice clean modern apartment, she sounded desperately unhappy and lonely. Notwithstanding this, eating at the day care service meant a lot to her and made her feel accepted.

On the 19 July 1999, Ms Ume, a life reconstruction advisor for older people (LRA), said about Ms Sato,

> She does not like her son-in-law. I did not hear it directly from her but the LSA told me so. Since she has established *ningen kankei dekiteirukara* (trusting human relationships), Ms Sato says such things to her. The other day, when she was *guchiru* (grumbling), she started to get angry and said, 'My daughter is wrong...she left me alone at this place!' She is always like that. She says, 'I hope a pick-up will come soon [pick-up from the other world, which means death]' But I think she was very happy that you, Otani-san, sat next to her and talked with her. She wants someone to listen to her grumbling.

After I said that I thought she was not grumbling about her son-in-law but may have had a headache, Ms Ume continued,

> She was *amae* (behaving like a spoiled child, coquettish). I do not think she had a headache. But it was her way of trying to get attention from others. If she says so, and says, 'I am leaving', then people come and ask her to stay. She has *shinpai-shō* (a habit of worrying about things). When she lived with her daughter, she went to her daughter at 3 am and woke her up and asked the same question repeatedly like she was worried about something. She is also *binkan* (sensitive). She cannot hear well but she can sense if people are talking about her. She may pretend that she is not hearing, but she senses it, sometimes to the extent of *higai-mōsō* (persecution mania). So we try not to talk about her in front of her.

Although she said that she had made the decision to move into the apartment, deep in her mind she might have blamed her daughter and probably her son-in-law too. She might have thought she would have been able to live happily with her daughter if it were not for the son-in-law. She said, 'My son-in-law is in hospital but he will come back home soon'. Nonetheless, from what others told me, she had more reasons for living alone than just her son-in-law.

I met Mrs Kawakami at a tea party at the Haruyama TSH, where I worked as a volunteer waitress. She was aged about 60 and had two sons in their 30s, both of them married. Her relationship with her eldest son stopped around the time of the earthquake. She did not explain why. Her youngest son had a two-year old son of his own, her only grandchild. While she enjoyed having their company, she was sad that her daughter-in-law spent more time with her own parents and did not bring her grandson to visit very often. She felt that it had become common for a son to be taken away by his wife's family. She said that she had wanted a daughter who would remain close to her, adding that I would be a good daughter because I sat and listened to her. Although she lived with her husband, he did not come to the tea-shop at the TSH *fureai* community center.

In this case Mrs Kawakami had lost her son to a daughter-in-law. There were also cases where an old woman had established a strong relationship with her daughter-in-law thinking of her as *kawaii* (a favourite), more so than her real daughter.

Another elderly woman who I met during a *yūai* home visit at the Natsu Aki PRH said, 'I am fine because I have a son. He comes to see me and brings food to me'. But as Mrs Sakura asked her more questions, the old woman started to reveal that her son did not visit her that often. Mrs Sakura later explained to other volunteers that she thought that the old woman's son was in his fifties and was too busy with work to visit often. It seems that old Japanese women still think that it's important to appear to have a good relationship with their children and might feel ashamed to make it known that they are not taken care of.

Most elderly people in Japan are strongly opposed to the idea of living with their children as they do not want to be a burden, even if they feel lonely living alone and apart from their children (Hashimoto, 1996). But physical separation did not mean emotional separation. The absence of meaningful relationships (Russell and Schofield, 1999) was more of a problem than living alone and many felt ashamed to say that they were left alone in TSH or PRH.

## Meaningful relationships with neighbors

Mr Yamada changed dramatically over the years following the earthquake. He had lived in the Haruyama TSH and later moved to the Natsuyama PRH. For the first two years in the Haruyama TSH he didn't speak to anyone.

About 50 years old, he was a car mechanic before the earthquake and had only five years schooling at the primary level. He did not receive further formal education because of the war. 25 years ago, when he was 28, he married a woman from his workplace, 16 years older than he was. He took care of her until she died of cancer and at the time of the earthquake he was a widower living alone. At the Haruyama TSH, when he saw an old woman in her seventies living near his shelter, he helped her to carry big items of furniture. Other neighbors gossiped that they were boyfriend and girlfriend, and accusation that the old woman denied. With the exception of such rare occasions, he did not interact with others.

After three years, he started to drop by at the tea-shop at the *fureai* center to meet with Mrs Ogawa, a JOCS[5] volunteer public health nurse. After his experiences in the earthquake, he lost trust in people as no one, not even his own family members, asked him how he was doing or tried to reach him. In the beginning, Mr Yamada would bring a plastic wrapper to donate as well as a clock. Because he did not believe in people's good will, his thinking was always give and take. However, after several visits he began to see himself as a customer. This pleased Mrs Ogawa very much as she saw it as a great step forward. Nonetheless, he always sat in a tatami room with volunteers from outside the Haruyama TSH; it took some time before he started to talk with other residents from the TSH.

After Mr Yamada won the lottery for an apartment in the Natsuyama PRH compound, he would still visit the tea-shop at the Haruyama TSH every week to see the JOCS volunteers. The Christian volunteers helped him to look positively on his new life and he volunteered as a representative of his apartment building in the self-governing body.

He had ambitious plans for generating community at the PRH. He said,

> I am thinking of holding a weekly tea party like this one at the community assembly hall of our PRH compound. I am thinking of volunteering as the accountant. I am thinking of becoming a *kanri-nin* (building manager).

The volunteers encouraged him, even suggesting that he volunteer to be chief of the self-governing body. He responded by saying '*Utsuwa de nai* (I know my limits)'.

On several occasions I saw Mr Yamada walking outside just to see if there was anyone to talk to. Sometimes he would even approach people. His facial expression had changed over the year since I met him in the TSH compound. He seemed to be no longer lonely.

## Human contact

There is an extensive gerontological literature on the problem of social isolation for elderly people. Russell and Schofield (1999: 69–91) define social isolation as an 'objective state of having minimal contact'.

Human contact is important for the development of community. It is also a measure of happiness and health. Human contact can be measured by the extent to which a person exchanges words with others. This is consistent with what I observed during my fieldwork.

Community activities prevent people from *tojikomori* (closing in oneself, not going out). *Tojikomori* is common in communities with a large elderly population and a high proportion of people living alone. In a NHK TV program broadcast on 17 January 1999, Professor Sachio Konishi of Kwansei Gakuin University said of residents of PRH,

> They were lonely at the beginning when they moved in, but they established relationships there in these three years. It was easier to form a community and relationships in the tenement house type residence with volunteer coming from outside. The question is if a community can be formed at public reconstruction housing.

On a News Park Kensai program broadcast on 17 January 1999, the Hyogo Prefecture Governor Kaihara said,

> I understand that it is a lonely time just after moving into new housing in a strange environment. We, the public administration, will support people. We will oversee community formation at these new places. We will continue to provide our support to real reconstruction.

To build connections, self-governing bodies allocated LSAs to visit residents living alone and to support community development. However, because they did not have enough staff to cover everyone, LRAs were employed by the city with the earthquake reconstruction fund to look after people not in general housing. One LRA was employed for every 100 people.

In some PRH communities, the LSAs took more active roles in organizing activities; in others they were less productive. Practical leadership was important for community development.

On 17 January 1999, NHK reported that earthquake damage was 'biggest in the area with a traditional community of human contact'.

Most fires occurred in these areas and the destruction was most severe. Reconstruction was also slowest there.

In an NHK program on 17 January 1999 about the Edagawa-cho TSH, Nishonimya City, Mrs Teruki Tanimura, *jichikai cho* (chief of self-governing body), said 'there is no human relationship of the back alley at public reconstruction housing'.

A number of TV programs in 1999 reported that elderly men and women experienced loneliness as a lack of human contact in PRH. On ABC TV, 13 January 1999, Mr Keisuke Kawabata (60) said, 'The days when I have no exchange of words with anyone continue'. On NHK, 14 January 1999, Mr Yasuo Uematsu (70) was reported to have had no human contact or *kōryu* (personal exchange).

On the 14 January 1999, NHK reported that Ms Fumiko Hirata (85) had moved into PRH six months ago but had no friends there. She was visited by the life advisor but would not leave her apartment and would only have brief contact with others. On the same station, 17 January 1999, it was reported that Ms Maki Nagano (85) planned to move into PRH in March. She said, 'I will be lonely. In public reconstruction housing, once you close the door, there is no human contact with the outside'. Also on NHK, 17 January 1999, Ms Toshie Murata, who had moved into PRH April 1998, said, 'Before this, I lived in a temporary shelter and I used to see people walking past the window. I had human contact with them. But now there is no exchange with my neighbor'.

The old people's home in Hochschild's (1973) study was built in such a way that it was almost impossible to walk from any apartment to the elevator without being seen from the living room windows that looked out onto the porch. People could glance up through the window and see or wave to neighbors as they passed by. This kind of housing design was effective for community generation and helped to prevent feelings of isolation among residents, although it did present a problem for privacy.

Mrs Sakura emphasized the importance of *tsunagari* (being related to other people) for community generation. She would attend *mimamori* meetings to report problems in this area.

## *Fureai*

Each TSH had a *fureai* center. *Fureai* can be translated as interaction, human contacts and making friends with new people. By conducting research in English, I found that I translated this word differently depending on the context or on what I was analyzing at the time.

Film director Mr Yoichi Yamada, who filmed the popular film series *Tora-san in downtown Kobe*, said in an interview with NHK, 17 January 1999,

> This Kobe City downtown area had warm human relationships. There was an intimate relationship between a shopkeeper and a customer. An old woman comes and shops every day. A shopkeeper says to her, 'You bought this yesterday. Why don't you buy this? This is a good deal today'. There is a relationship. That relationship was seen in the traditional shopping streets. This is the core of daily life.

In modern PRH built after the earthquake, there were no back-alley shopping streets. Instead, there was a huge American style supermarket near the train station. Elderly people need to have a place for human interaction as part of their daily routine.

A NHK report on 13 June 1999 explained this situation:

> We try to report on the residents living in collective housing, those seeking a life style with *fureai* (interaction with other people). The old people who have repeatedly moved from their old houses, evacuation centers, temporary shelter and then to new permanent housing are no longer good at *fureai* (making friends with new people).

This suggests that *fureai* is important for reducing loneliness.

Volunteers were also an important source of human contact for elderly people living alone. When a volunteer brought a resident a hot meal, they showed that they cared. Volunteers can do this more convincingly than bureaucrats, public officials and civil servants (Vaughan and Buss, 1998).

Town planning must take into account aspects of community life that enhance the quality of daily human contact, especially in ageing urban societies. Social interaction helps to makes life worth living and is therefore important for healthy ageing (Wilson, 2000: 107).

## *Ikigai*

*Ikigai*[6] (value, meaning of life) is important for being happy and not lonely. If a person has things to do and feels useful to society, regardless of whether it is paid work or not, they can feel that life is worth living (Gurley et al., 1996). Some people lost their *ikigai* in the earthquake and gave up finding another. Others who did not have one found that the earthquake reconstruction created new opportunities for them. It is possible to find a new *ikigai* if an old one is lost.

In a NHK earthquake special broadcast 15 January 2000, it was reported that Ueda and Haruko Kuroda were helping to create jobs in a disaster affected area by preparing for a restaurant opening. Ms Kuroda said,

> [The disaster affected] *jibunde* (produce the goods by themselves). They sell what they produce by themselves and if they can feel that they are earning by what they have produced, then this becomes *ikigai* .

In another example, reported on by Mainichi TV on 17 January 1999,

> Mr Yamamoto is disabled. He used to complain a lot. He did not have a purpose in life. Now he has started to thank the earthquake which shook his life and woke him up. Since the earthquake, he has been active in a community support group. He is positive. He feels he has a purpose in life.

Disability need not stop a person from achieving meaning in their life. In an interview on 22 July 1999, the director of an old people's home in Ashiya said,

> The experience of group housing made me realize the importance of a mixture of different kinds of disability. Those with different abilities left can support one another. When they can be of help to others, it becomes their *ikigai*.

Ms Kazue Tanii's (89) has not found meaning in her life. After she moved into PRH, her daughter, Makiko, died. She told Sun-TV on 10 November 1998, 'I sometimes think it is better for me to go to heaven. I am no use to anyone'. Also, a 60 year-old alcoholic man at the Haruyama TSH died of malnutrition. The loss of his job and his *ikigai* could have contributed to his condition.

An important aspect of personal meaning is what Giddens (1991: 36) has called 'ontological security', a person's sense of order and continuity in relation to the events in which they participate, and the experiences they have, in their day-to-day lives (Giddens, 1990, 1991, Mellor 1993: 12). In this sense, *ikigai* can be derived from a 'practical consciousness' of the meaningfulness of our day-to-day actions (Mellor, 1993: 12).

## Gender

The demographic and socio-economic data show that the study population in TSH and PRH belongs to the most socio-economically disadvantaged class with the largest proportion of elderly people, mainly women. This disadvantage was often referred to by the media and by public officials as a problem of elderly women specifically. Yet public officials, media coverage, and other key informants said that women are coping better than men, despite their economically disadvantaged situation.

The Director of the Department of Health and Welfare said in an interview on 26 January 1999,

> Women are doing well. The problem is older men living alone. They lose their vitality and motivation. If my wife dies, imagine, I will have to worry about how to become myself.

There are men who are social no matter whether they are single or not. In Japan's modern urbanized society, men often make friends through work even if it means they could not afford time for their family and community. These men often depend on their wives to make friends with neighbors. This is why elderly men living alone can become isolated more easily than women.

Women, on the other hand, have a more varied link to decision making processes and this is related to their level of happiness. Participation in the evaluation and planning of resource allocations can empower women. Some women choose to be quiet and passive. They think perseverance is a better coping strategy to survive old age and to maintain their security (Hashimoto, 1996).

I observed in my fieldwork that women found it easier to make friends in general. Making friends is a major means of preventing loneliness and helps a person to have a healthy life and to eat well. The gender issues surrounding the ability to make friends are important for understanding the dynamics of community development.

It was common to here talk about the 'feminization' of the TSH and PRH populations. A high proportion of residents were women and they were often seen as weak. A similar pattern can be observed in old people's homes where only one out of ten residents are male. This may make things difficult for men, as they are generally in the minority.

Media and high-ranking government officials often referred to the residents of TSH and PRH as *obah-chan* (grandmothers, old women). However, this was not the case and many men lived there too.

It is easier to sit and talk to women and help them to participate in activities than it is with men. Mrs Sakura suggested that elderly men needed to be treated with a special tactfulness. She said, ‘The key to get a man talking is to talk about sports. Talk about baseball. Find topics that interest him. When we talk about baseball, any men can join in’. An old man at a group home used to stay in his room as he was not good at chatting with his neighbors who were mostly women. The man started to do odd jobs and some carpentry work. He would go to the common room to find out what he could do for the community. His work made him happy because he felt that he was useful.

A NHK program broadcast on 14 January 1999 reported on Mr Yoshiaki Murai (70), an old man living alone. He would leave his iron door open on purpose so that people might drop by to say hello. He said,

> I can have a chat with a [female] neighbor, but I cannot invite them into my apartment. No matter how old she is, a woman is a woman. It would be a problem.

He teaches singing once a week in a common room. He goes out to a café to make friends. There is a café where older men get together to chat, one of the few opportunities for men to make friends.

Special efforts needed to be made to bring men into the community generation process.

## Summary and conclusions

In my fieldwork observations and in the media coverage there were many cases of TSH and PRH residents saying, ‘I am lonely’. They would say, ‘I cannot make friends at PRH’, ‘I miss TSH’, ’I am alone and I am lonely’, ‘I am old and I cannot make friends’ and ‘I am a man and I cannot make friends while women make friends more easily’. Loneliness is one of the most important challenges of life in TSH and PRH, but the media tended to blame examples of loneliness on the failure of the public administration.

A community is a group of reliable and helpful people who stay together in times of success and failure. PRH was a place where hundreds of people were lonely together. TSH was also a place where hundreds of people were lonely together but they shared a similar experience of loss and the setting was more open to exchange. Residents said a big city was a lonely place, but the city alone did not create loneliness. Instead, loneliness resulted from the lack of a close and familiar community.

PRH was a lonely place for the residents; there were many people around but they did not exchange words. Many residents had left communities in TSH. If people had a choice in their place to move to, and if they had been allowed to move together as a group of neighbors, they might have felt less lonely.

Natsu Aki PRH and Fuyuyama PRH were in the process of building dynamic communities, but most other PRHs had no such dynamic, sometimes resulting in *tojikomori* (residents locked in their apartment). Building PRH community was important for providing residents with a sense of *ikigai*.

According to Vanier (1989), remedies for loneliness are mutual affection and participation in a genuine community. This needs the individual to have active and positive attitudes towards their community and to work with others rather than wait passively for someone else to make things better. Some people are lonely wherever they go; they focus on what they do not have and on what others have, and what they used to have but have lost. Some people are happy as they know how to be content and how to be happy whatever they have; they have hope. There are therefore individual issues and social issues which contribute to loneliness.

According to the media, the health of elderly residents worsened during their time in remote rural TSH. However, during my fieldwork I observed that life in rural areas had benefited residents as well, especially after the development of community. It was difficult for many residents to leave their community when their TSH compound was closed.

Although the physical health of residents was the main concern in the first years after the earthquake, as time progressed mental health became more of a concern. Loneliness was seen as a major difficulty following from the numerous relocations individuals faced after the earthquake. *Kokoro no* care (emotional care) and *tsunagari* (being related to someone) became key words for activities and policy. A common name for a community center was the '*fureai* (human contacts) center'.

Special attention has been paid to providing care to elderly people in the aftermath of a natural disaster, but there are still unresolved problems. One of the main challenges is how to ensure that disaster-affected people are able to build human relations in their new living environments.

# *Chapter Eight*
# Kodokushi

On 6 January 2002, at the beginning of the seventh anniversary of the earthquake, the front page headline of the *Asashi* newspaper was *kodokushi* (dying alone). The article said that more than 40% of PRH residents were aged over 65 and 30% were living alone. This meant that *kodokushi* was a cause for concern. *Kodokushi* was a buzzword used frequently in media coverage, in newspaper headlines and in the titles of special TV features.

Most *kodokushi* reports were cases of deaths of elderly people. Research conducted by Doshisha University in 1997 found that most of those who die alone are men in their 50s and women aged over 70. This chapter looks at dying alone in the Japanese context. It asks if this is a new phenomenon in Japan and if it is unique to Japanese society. It also considers if cases of dying alone can be prevented and how society should respond. It suggests that in many cases *kodokushi* eventuates due to an extraordinary absence of family.

## Headline grabbing

*Kodokushi* is a sad word. It was used by the media to create attention-grabbing headlines, often implying that the public administration was to blame for the lonely deaths of elderly people living alone. It was not commonly used before the earthquake, but afterwards it was used frequently. In Japanese society, if a person has no one to care for them when they are old and frail and they die alone, it is often seen as an indication that they had not lived a good life.

*Kodokushi* brings two fears together: isolation and death. Due to sensational media reporting, it has also become associated with the earthquake, especially with the condition of elderly PRH residents.

While *kodokushi* is often used as a sensational newspaper headline, on TV news it is difficult to get the same effect as an image of a dead body cannot be shown on TV. When a person is found dead, the ambulance service is called, not a TV reporter. TV news can interview people about the *kodokushi* case, but TV cannot show the moment of a person's death. Nonetheless, although *kodokushi* is a very private event, because it is linked to the earthquake it is a public issue.

The word *kodokushi* causes tension when it is used. Although it often sounds shocking, it also expresses grief by drawing the public's attention to the sad event of an elderly person's lonely death. Media reporting on *kodokushi* cases can be defended if it helps the public to mourn, even though it was not able to prevent the person from dying alone. Society should feel deeply for people who die in this way (Walter et al., 2000: 22).

Although the media reports frequently on *kodokushi*, it does not report on particular cases. The deceased individual's story would itself not be newsworthy, but more general reporting about *kodokushi* is.

During an interview on 26 January 1999, the Director of the Department of Health and Welfare of Kobe City said,

> The other day when we had a meeting, [Ms Takada] told me that she was annoyed by the way the media reported a recent case of *kodokushi* at the temporary shelter there. Her group had been aware that the person was at risk. They had been visiting the person everyday and paid particular attention. But the person died when no one attended. Then the media reported loudly, 'Another *kodokushi*'. She said that it was not *kodokushi*. No one might have attended at the moment of the person's death but the volunteer group had been making a lot of effort to attend the terminal stage of the person.

Ms Takada was a nurse and the leader of volunteer groups working at the Seventh Seishin TSH. It was possible that Ms Takada sees actual *kodokushi* cases as arising from the failure of NGO efforts to foster a caring community at TSH and that the director had felt threatened by the critique of *gyōsei* (public administration) policy.

In an interview on 22 July 1999, the director of an old people's home in an affluent area of Ashiya city said,

> *Kodokushi* has been an issue at the disaster-affected area. *Gyōsei* told us not to poke our nose into it because they fear that [residents] would be dependent and would be trouble later. I believe it is not good to leave them alone. Appropriate care is necessary and it will help [them to be] independent.

*Gyōsei* were sensitive to criticisms that held them responsible for preventing the incidence of *kodokushi*. They claimed that they provided enough assistance to residents of TSH and PRH and saw *kodokushi* as an individual problem.

## Earthquake vocabularies

Alongside other health measures, *kodokushi* was used to assess the effectiveness of community care services long before the Hanshin earthquake. Japan's ageing population first came to the public's attention in the 1970s, particularly concerning the health and loneliness problems of elderly people living alone. Okamoto (1994) said that elderly people who were accustomed to living in big families were facing various problems in the rapid change to nuclear families. These problems included loss of order in daily life, poor health, loneliness and suicide. She argued that these issues have made individual problems like *kodokushi* appear as social problems (Okamoto, 1994: 33). She proposed that community care could lighten the sense of loneliness and provide elderly people with something to look forward to in their daily lives. She reviewed the activities of the daily food service, an example of community care, from 1972 to 1992.

Before the earthquake, as many as 150 cases of *kodokushi* per year were reported in Kobe City (Ueno, 1997: 152, Takayose, 1999: 92). It started to receive public attention when cases were reported in TSH, bringing to the surface previously unseen problems of health and welfare among elderly people (Ueno, 1997: 152, Takayose, 1999: 92).

On 17 January 1999, NHK reported that,

> Volunteer work [supports] the life of older people. As the life recovery is prolonged, the role of the volunteer changes. For those older people who have no plan on how to reconstruct their life, a volunteer who just sits and talks with them takes on a critically important role.

In this program, *kodokushi* was discussed in the context of the role of volunteers in supporting elderly people and community activity as an opportunity to develop *tsunagari* (relationships).

It was common for the media to talk about *tsunagari* after the earthquake. It also became common in daily speech in the Kobe and Osaka areas. Other words that entered the vocabulary at this time include community, volunteer, *kodokushi*, *seikatsu fukkō* (life reconstruction), *fureai* (human contacts), *kōryu* (exchange) and the iron door.

## Anxieties about the future

A report prepared by the Citizen and NGO Disaster Prevention International Forum Executing Committee in 1998 listed three issues exposed by the earthquake: 1) the lack of preparation for the ageing society, 2) *kodokushi* and other risks common to elderly people and 3) weakening forms of traditional community and family support in the context of ineffective social security. Together, these constitute the key anxieties about the future of Japanese society.

The report argued that men in their 50s to early 60s who lived alone, without family, work or *ikigai* (life meaning), are at great risk of alcohol dependency and *kodokushi*. For governments to reduce this risk, the provision of employment and means to discover *ikigai* are important, in addition to sustained community support through initiatives like establishing *kokoro no kea* (mental care) centers and home visit programs involving public health nurses and a volunteers. The report recommended that people with alcohol dependency should be given timely professional treatment by doctors, the public social security service and support groups and opportunities for human contact through community activities should be increased.

An article in the *Asashi Shinbun* newspaper, 9 January 2000, reported that 38 PRH residents had succumbed to isolated death or suicide in the previous year. It said,

> At a temporary shelter, *kodokushi*, which counted over 230, became a social issue. This remains the issue at public reconstruction housing. The proportion of old people at public reconstruction housings is high and it has been pointed out that this is the miniature model of the future Japan's ageing society. Looking at the sudden death of their neighbors, the residents are seeking a way to help each other to live through the community development such as the formation of the self-governing body.
>
> Prof. Yuzo Okamoto, Kobe City Nursing College (Gerontology) said, 'The residents can help each other among neighbors, but they cannot be responsible to stay with one until one dies. The death of those living alone at the public reconstruction housing shows, through the life there, the universal policy and administration issues relating to the ageing society that we will face soon. The *gyōsei* has to grasp precisely the needs of each individual who is living alone and needs to speed up to provide the support to pull out the desire and spirit of the residents to be active'.

The reported cases of *kodokushi* and suicide at the Seventh Seishin TSH created a sense of crisis and resulted in the recognition of the need for community activities such as *yūai* home visits to old people living alone (Sugama, 1999: 311). The home visit was traditionally carried out by *minsei iin* but in Seventh Seishin TSH and other smaller TSH communities they extended them to a women's group and other volunteer groups (Sugama, 1999: 312). This was later extended into PRH communities, including my fieldwork sites Haruyama TSH and the Natsu Aki PRH.

Doshisha University in Kyoto conducted research on the people vulnerable to *kodokushi* and published its report in October 1997. Although the title of the research was *Kodokushi*, it did not study those people who actually died of *kodokushi*, but rather the population at risk of *kodokushi*. It was an empirical study of the health status of *sōnenki* (people in their 50s to early 60s) in the context of the Japanese welfare state. The researchers interviewed 169 TSH households from 25 January to 26 March 1997. The people they interviewed were mainly old people living alone, disabled people and *tojikomori* (people who did not socialize with others).

The research saw *kodokushi* as a product of broader social problems. The fear of *kodokushi* reflected various anxieties about future life including financial matters, housing and the maintenance of close human relationships. The report surveyed the views of TSH residents about the *fureai* (human contacts) community center and volunteer workers. It suggests that some see *kodokushi* as a failing of the self-governing body and volunteer workers who are responsible for increasing *fureai* among the residents, especially elderly people living alone.

The researchers argue that the problems faced by TSH residents reflect widening class differences due to labor migration and urbanization along with the focus of economic growth and development policy in concentrated areas. This broader critique focuses on the shift by the central administration to a welfare state model based on self-help. It argues that government uses population ageing as an excuse for the failings of this model.

According to the research, people in their 50s and early 60s are in as much need of support as those aged 65 and over. The age line drawn in the social welfare system means that they are not entitled to any pensions and do not receive the free public services that are available to silver citizens.

The research also outlined how the TSH had successfully mobilized the community to eliminate cases of *kodokushi* through slogans like 'Don't let the old person living alone *kodokushi*' and 'Do not let the old and disabled people be bedridden'.

## *Kodokushi* and the lack of family

In Japanese values, elderly people living alone receive special attention and support. The idea that elderly people should not live alone is also deeply entrenched in Japanese values. This idea motivates a great deal of the research on the post-earthquake reconstruction. However, the number of elderly people living alone will increase in the future and this it is not necessarily true that a person living alone is lonely or isolated from their family.

There was not a strong focus on family in the media coverage of the earthquake. In 2000, the media did not refer to family at all. Instead, the focus was how to establish community care in ageing community as well as in society in general. There seemed to be an assumption that elderly people have no family support.

Immediately after the earthquake, media references to family were focused on cases of loss. For example, when a Chinese restaurant owner was rescued, he said to the Yomiuri media, ‘I am OK. Please rescue my family first’. Sadly, he was the only survivor found; his wife and children all died. He says, ’I want to start a family again. Without family, I do not know why I am working hard’. These are examples of how drastic sudden loss can happen to a family support structure, but this is not the focus of this research.

In a NHK report, 11 January 2000, an 80 year old woman who lived alone in PRH thought her dead daughter was her guardian angel. Although the report did not say how her daughter died, it is possible that her death was caused by the earthquake. She said while looking at a picture of her daughter, ‘I do not know why only I survived. My daughter is protecting me’.

In a case reported by Yomiuri, 12 January 1999, an 87 year old father and his 50 year old son committed suicide together because they were tired of living in TSH. Although they had won a lottery allocation for PRH, they could not afford the rent because the son had a debt of 1,100,000 JPY (about £5,000) from long ago. His father said it would be better for him to die than to go to an old people’s home. This is a more radical case as both of them killed themselves.

When adult children are not married, they tend to live with and take care of their elderly parents. When adult children have their own families, their parents tend not to expect them to cohabitate with them. But there are exceptions. In one case, reported on NHK on 10 January 1999, a mother lost her husband in the earthquake. After her daughter got married and had a baby, they all lived together. In these cases, the grandmother is often needed to take care of the baby so the mother can work.

In another case, reported on Yomiuri, 24 November 1998, an elderly couple lived with their son and his family for a few months after the earthquake but moved to TSH and then on to PRH. They did not want to be a burden to their adult children and their families.

During my fieldwork, some elderly people mentioned that they had tried to live with their children, but had to move to PRH because they did not want to be a burden. In some cases they had experienced friction with their children or in-laws. Some were quite happy to live separately, some others would initially say that living alone was their choice, but after a time they would admit they were unhappy or even angry about their situation. Some felt that their children were stolen by their in-laws.

Once children have started their own family, elderly people might have to establish a social network even though their family remains important. In general, spouses and members of the immediate family provide the most support in times of crisis and relatives, particularly daughters, are more likely than neighbors to provide long-term assistance (Bowling and Browne, 1991). When no family is available, it can lead to *kodokushi.*

In TSH and PRH, the emphasis was on elderly people, especially those living alone. The media would occasionally introduce husbands and wives who lived together, but it was very rare to talk about children. The focus was on community support and development.

In one case reported on NHK, 11 January 2000, 86 year old Nakano was said to take care of his 76 year old wife who was recovering from temporary dementia triggered by the earthquake. They had three children and six grandchildren, but the report did not focus on their relationship. Nakano said he had experienced the Kobe flood, the bombing of Kobe during the war and going to war himself, so 'living in an evacuation center was not a great challenge'. He was, however, concerned for his wife. This case showed how the expectations of elderly people of their family were not the same as they used to be. Children and grandchildren may provide emotional support and *ikigai*; watching children and grandchildren grow up was often a source of pleasure for elderly people. However, children were not expected to provide daily practical or financial support.

In some cases, children and their families visited their elderly parents and brought food etc. They explained and wrote down instructions on, for example, how to use the modern bath at PRH, which button to press and how to adjust the temperature and so on.

The director of an old people's home in Ashiya described family life as follows:

> People dine together, watch TV together in a common space. A mentally disabled person shows a great improvement. Those who have been neglected and excluded, they can have a family-like life here. It is very homely. A psychiatric doctor has been surprised to see such change when they come here.

This is family life in the sense that members spend time together eat together, watch TV together and keep each other company. There is not necessarily a bread winner, earning income to support family members.

In short, the media did not emphasize family as a main issue in the happiness of elderly people living in TSH and PRH. When it did talk about family, it was not in the extended, traditional sense.

## Community change

On 9 January 2000, an *Asashi* newspaper headline reported that there had been 38 cases of *kodokushi* or suicide at PRH in the past year, in addition to the 230 cases of *kodokushi* in TSH. It argued that these numbers made *kodokushi* a social issue. It showed a photo of *mochitsuki* (outdoor rice cooking to prepare for the New Year) in which old people were making *mochi* together. At one temporary shelter, *kodokushi* counted over 230 and became a social issue:

> Looking at the sudden death of their neighbors, the residents are seeking a way to help each other to live through the community development such as the formation of the *jichikai* (self-governing body). The *jichikai* is critical to prevent the incidence of *kodokushi*.

The presence of a *jichikai* showed that a community was active and that they had activities where neighbors could meet and develop relationships. It also showed that the residents paid attention to neighbors in need or that the community had mechanisms to support weaker members:

> Nobody died without other people knowing. Gichi Igata, a 92 year old man who had lived alone before the earthquake, found himself livelier in the temporary house than before the quake.
>
> The Great Hanshin Earthquake shed light on the solitary life of elderly people living in urban areas. Among the victims of the quake, more than 30 people died or committed suicide without the knowledge of any other people, even after they were moved from temporary homes to more permanent disaster-relief public housing.

These newspaper articles see *kodokushi* as a symbolic phenomenon which shows the failure to develop a caring community.

Although a *jichikai* was critical to prevent the incidence of isolated death, *Asashi Shinbun* newspaper reported on 9 January 2000 that about 20% of

PRHs were yet to form one. The formation of each *jichikai* was the core of local community, but their establishment was not going as smoothly as expected.

The organization of residents who administer the self-governing body and *danchi* (housings) had played a critical role in the rescue activity immediately after the earthquake and during the post-earthquake reconstruction. At the TSH, the incidence of isolated deaths triggered the formation of the self-governing bodies.

According to *Asashi Shinbun* newspaper, 9 January 2000, among the 220 PRHs in Hyogo Prefecture, 180 reported that they had formed a self-governing body.[1] Those who had not yet formed one were mainly in the urban areas such as Kobe City and Akashi City. Some self-governing bodies were established but then fell apart because of disputes among the residents. This showed the difficulty of running a new community which was formed rapidly in a short period of time.

There are three possible reasons why the formation of self-governing bodies had not gone smoothly: 1) Many of the residents were too old to participate in the activities of the self-governing body, 2) They were used to receiving support and subsidies from volunteers and *gyōsei* (public administration) at TSH, so they were not accustomed to being *jishu soshiki* (independent and autonomous), and 3) Many of the residents of PRH were not affluent and could not afford to take care of others. Also, because PRH were built in suburban areas, it was difficult for family members to make frequent visits and contribute to developing the community.

On 13 January 1999, Super News Kansai reported that Mr Tetsuo Matsumoto, an LSA, had found an 82 year old woman who had died alone. He found her body a few days after her death. In response, Mr Hiroshi Hayashida, Division Chief of Kobe City Home Welfare was reported to have said,

> We need to create a society where old people can take care of each other. It is important to make a good community. With many old people who are no longer active, it is difficult to form an active community.

The report added,

> The public reconstruction housings are a highly aged society, but this will soon be an issue in all of Japan, not only the disaster-affected area. We need to strengthen the public administration system and the local community to support the old people's self-support life style.

In this report, *kodokushi* was seen as a result of the failure of *gyōsei* to collaborate with residents to develop a caring community.

This view was shared by the *Kodokushi* Lawyers Association. They said in a submission to the Hyogo Prefecture and Cities that, 'Public administration also has responsibility for the isolated deaths which are occurring among the disaster affected people'. Another group, the NGO Nishinomiya Local Help Network added, 'The public financial aid will end by the end of March. Each NGO has to find a way of fund raising to continue their activity', drawing attention to the view that local authorities were ineffective in encouraging the development of community. The media was calling for public financial support for NGOs so that they could continue their activities. They stressed the importance of the work of NGOs in the prevention of *kodokushi*. The media saw *kodokushi* as a failure of *gyōsei* and their lack of support for the work of NGOs.

In a NHK special broadcast 17 January 1999, the panelist Prof. Sachio Konishi of Kwansei Gakuin University said,

> A temporary shelter is a temporary shelter. It is too cold in winter. It is too hot in summer. They were lonely at the beginning when they moved in. But they established those relationships in the tenement house type residence and with outside volunteers. The question is if a community can be formed at public reconstruction housing.

He was responding to cases of people who said that TSH life was better. He said community was easier to develop in the low-rise apartments like those that some had lived in before the earthquake in downtown areas. Low-rise was also the design of TSH but not of PRH. He also pointed out the role of volunteers in the process of community development. He suggested that these insights could be taken into account for the new challenge of community development in PRH.

On the same TV program, Ms Toshie Murata, a resident of Silver Housing, argued that the TSH design was better because it had a window through which she could see her neighbors walk by. The potential for interaction arising from this is what Hochschild (1973) called 'unexpected community'. At high-rise PRH, she had no such opportunity to interact with her neighbors. Although her Silver Housing apartment was on the ground floor, she couldn't see people through her window. Instead of going outside, PRH residents often *tojikomori* (stay closed at home). Often they had no friends or volunteer visitors.

In the United States, where elders are often housed in tower blocks, broken lifts can lead to serious isolation, depression and inability to leave the home (Wilson, 2000: 106). In Kobe, even when the lift was not broken, *tojikomori* became an issue among elderly PRH residents. The large size of apartments meant that there was a greater distance between neighbors and thick walls and heavy doors meant that, unlike at TSH, residents could not hear their neighbors.

In a NHK program, 17 January 1999, Professor Murosaki of Kobe University said, 'Support from community and neighborhoods means a lot to older people. These are broken up every time they move. It is difficult for an old person to start from the beginning, to adjust to a new environment. We must think of housing not only as a container. We have to look at reconstruction from economic and employment perspectives.'

According to the media, public and community attention should have prevented cases of *kodokushi*. At TSH, according to *gyōsei*, the support of emergency volunteers and public health nurses would have also improved the situation. In PRH, this support wasn't available because *gyōsei* considered the emergency period to be over. Nonetheless, PRH residents were still affected by the earthquake and cases of *kodokushi* should be seen in relation to this.

According to Hirayama (2000), circumstances at TSH were 'thoroughly effective in causing the dwellers to become depressed'. He said that lone deaths[2] together with suicides and accidental deaths were increasing problems.

By the media reports, it seemed that TSH residents were able to form, although with difficulty, communities similar to what they used to have before the earthquake, but could not do so again once they moved to PRH. This was not only because of the high rise design of PRH. It was difficult for many of the elderly people to make new friends and many may not have expected their neighbors to live very long. Residents might also have been discouraged when volunteer services that they enjoyed in TSH were not available in PRH.

## Feelings about old age

Most reported cases of *kodokushi* were of men in their 50s and 60s and women in their 70s and 80s. Some of them were suspected suicides. The media and many unhappy elderly residents said that this demonstrated that those who had died had been mistreated. This generation of elderly people lived through the war and worked hard to develop Japan into a developed country. Many believe they should be cared for by the younger generation who were born into the affluent society that they helped to create. The Japanese Health and Welfare system provides elderly people with free access to medical care (a severe strain on the stagnant Japanese economy) and generous pensions and special benefits for war widows. *Kodokushi* among elderly people is not acceptable in Japanese values.

The number of *kodokushi* and suicide cases in TSH often appeared in news headlines. For example, on 15 January 1999, Mainishi TV ran the following story on the Seventh Seishin TSH:

> Mr Mamoru Noguchi (70), chief of the self-governing body. 74 people died in this temporary shelter in the past three years. Two suicides. Three *kodokushi.*

As this headline illustrates, post-earthquake *kodokushi* was often seen in relation to other issues such as poor health and unhappiness among elderly people and the economic recovery and stagnation in Kobe.

Elderly Japanese people fear isolation. The current generation of elderly people has retired from a life-long employment system. Typically, they had worked for just one company. Japanese retirees usually found another job for about a decade after their retirement. Not many enjoy retirement in the Western sense. Japanese society was workplace centered and men especially faced isolation when they retired or lost their job.

Those who had survived the earthquake might feel that they had died socially, especially those in their 50s who were not ready to retire but lost their job and workplace in the earthquake. Some might have lost the place of their activities which gave them meaning in life. Others might have lost their *ibasho* (place which identifies their relationship with society).

## Feelings about death

*Kodokushi* implies that a sick and dying person had no one to take care of them. It implies that no one cared about the dying person, or that their children were neglectful. To say, '*Oyano shinime ni aenai* (One cannot attend when one's parent dies)' is seen to be a most unfortunate thing in Japanese society. *Kodokushi* shows how perceptions of the breakdown of the family reflect the social changes currently occurring in Japan, such as population ageing, urbanization and so on.

*Kodokushi* was used by newspapers to shock and attract reader's attention. On TV news, it featured in headlines late in the 1999 wave but by the end of 2000 it was less common.

## *Good and bad death*

There is a Japanese proverb that is sometimes told to a person who has done a bad thing: 'You may not be able to see your parents when they are dying'. It is seen to be very important to attend a parent's death; not attending can lead to profound feelings of regret.

In Britain, the 'perfect' death is one in which the dying person is surrounded by their family at home (Bradbury, 1999: 147). In this way, the dying person is able to say goodbye to their loved ones. In Japan, *kodokushi* is the opposite of a 'perfect' death.

Had the earthquake not occurred, it is difficult to say if the expectations of elderly people in Kobe about death would be different. Most of them lived downtown in low rise *nagaya* type housing. Many elderly people had neighbors who would look after them if they did not have a family. Some would have died alone anyway, only their deaths would have been less visible without the media focus on TSH and PRH.

*Kodokushi* is a bad death, wherever it occurs. No one wants to die alone. Old people interviewed on TV said they were scared to live at PRH alone because of the prospect of *kodokushi*.

In contrast to *kodokushi*, *pokkuri* (dying suddenly and with no pain) is typically seen as a good way of dying. Japanese people often say, 'When I die, I want to die *pokkuri*, then I would not suffer from pain and I would not have to be a burden on others who must take care of me during my death'.

In Britain, a hospitalized death, where a dying patient is unconscious and free of pain, is seen as a good death (Bradury, 2000: 61). A bad death, by contrast, involves a struggle for control.

If a person is suffering or is a burden to others, it is often thought that they should not live long. But a person who lives to an old age after raising their children and seeing their grandchildren is typically seen to have lived a good life. In Japan and Korea, people still often rush to get married if their parents become ill so that their parents can feel that their children are secure. A daughter or daughter-in-law might regret not delivering a baby before the grandparents died.

If a person dies young, especially if it is an unexpected and premature death, it is often said that they are too young and their life was just starting. Another saying, '*Tenju wo matto suru* (If you lived long, you have done your best and completed your duty and turn. No one can expect you to have lived longer)' suggests that in Japanese society long life is typically seen as a good thing.

From another point of view, it can be said that death by natural causes is a good death whereas an unexpected death by accident or carelessness is a bad death. If one dies as result of carelessness, it is often held that the person did not take care of their life or fulfill their duty to society. However, if a person dies without pain and has not been a burden on others, they are considered to have had a better death than a person who after a prolonged life has incurred expensive medical costs and worn out family carers. A person might want to die with dignity rather than prolong their life with drugs and modern medical equipment, especially if their life is not meaningful.

Most of the time, the subject of death is taboo. People respond to thoughts of death with fear. However, death can also be welcomed either as a peaceful end to suffering or an as an event of mastery over the challenges of life. Death is also often the source of attraction to religion.

In the modern world, family members who used to care for the dying now put them in the care of a public hospital (Walter et al., 2000: 15). Family members may attend at the last moment, but nurses and other workers are the ones who take care of the dying.

## Summary and conclusions

*Kodokushi* (isolated death) was a word that came to prominence during the post-earthquake reconstruction. It generally referred to the isolated death of elderly TSH and PRH residents who had no family to look after them. It was commonly used in news headlines to produce alarm and draw the public's attention. It was also the topic of sociological research on older residents of TSH. Most *kodokushi* cases were reported of men in their 50s and 60s and women in their 70s and 80s.

The image of isolated death is shocking and negative. It draws attention to social problems produced by the broad changes in Japanese society, including population ageing, economic stagnation, urbanization and modernization with the loss of the sense of local community. It highlights anxieties about the future and reflects feelings about community change, old age, and death in modern Japanese society.

If a person dies alone, it is thought that they did not live a good life. But, like in America and Britain, elderly parents often want their children to be independent. Nonetheless, in Japan, it is considered very important for a person to attend the death of their parent.

# *Conclusion*

This research was built on the notion that Japan is experiencing rapid population ageing along with other social changes. It found that the current system is unable to cope with the changes. There is a need for greater knowledge and new strategies for meeting present challenges. Focusing on the public reconstruction after the 1995 Great Hanshin Earthquake, the research drew on case studies of media reports, public surveys, original fieldwork and interview notes.

The recovery process after the Hanshin earthquake presented opportunities for experimenting with community development in a society comprised of large proportions of elderly people who are isolated from their families. Such people make up an increasingly important group as the populations of societies age. The post-earthquake reconstruction of Kobe was therefore a model of Japanese society in the future.

Specifically, this research looked at community development among elderly residents of high-rise PRH and hutted TSH. Community development in aged societies raises more policy and operational questions than could be considered in the scope of this book. Nonetheless, key lessons can be drawn from the exploratory micro-study methodology employed in this research.

How far is this research representative of Japan's future? As the numbers of elderly people living alone increase in proportion to the total Japanese population, so too does the public concern over the challenges to social security that this increase will bring. In demographic terms, it is said that the proportion of people aged 65 and above, around 40% of TSH and PRH in Kobe, will also be around 40% of the total Japanese population in 2040.

Elderly people who live alone have received special attention. Elderly Japanese women living alone have increased from 690,000 in 1980 to 2.29 million in 2000. Elderly men living alone have increased from 180,000 in 1980 to 740,000 in 2000. These figures are expected to increase to 3.81 million women and 1.76 million men in 2020 (Fukushi channel, 2002).

The population of this research was displaced suddenly as part of the post-earthquake reconstruction process. Future population ageing will be more gradual.

PRH communities can be said to be developing along the lines of traditional Japanese communities, but with new strategies to meet the challenges presented by high-rise apartment buildings inhabited by elderly people living alone. In PRH communities, I observed activities like those described in the national report on the system that supports the *jichikai* (self-governing body) for elderly people in 2001 (Ashita no nihon wo tsukuru kyokai, 2001), such as lunch services for elderly residents. Key groups and community workers were also the same, such as *minsei iin* (welfare commissioners), public health nurses, nutritionists and *jichikai* members. One of the priority areas of *jichikai* work was providing care and *ikigai* (meaning of life) to elderly residents. *Minsei iin* played an important role in providing advice to *jichikai*; 86.3% of *jichikai* received advice from *minsei iin*. The report concluded that communities with *jichikai* who received advice from *minsei iin* had more active and collaborative communities.

The report also concluded that the small communities surveyed had no capacity to cope with the increasing demands of an ageing population, and that it was increasingly important to coordinate between *jichikai*s for service provision. Mrs Sakura, a female welfare commissioner at the Natsu Aki area whom I met during my fieldwork, made similar comments.

The report also found that *jichikai* comprised of *yakuin*s (committee members) of both sexes have a higher rate of partnership with other groups, understand the community situation better and performs more active, better neighboring. The report, therefore, recommended that *jichikai* have *yakuin*s of both men and women, not predominantly men.

## Critical findings

### *Elderly people made visible*

In terms of morbidity and mortality, more than half of the immediate victims of the earthquake were elderly people, measured as the number of deaths, casualties and earthquake related illnesses. Medical professionals subsequently drew attention to the needs of elderly people (Tanida, 1996). Government prioritized the care of people aged 65 and older, especially those living alone, as well as disabled people. Both the Department of Health of Kobe City and Governor of the Hyogo Prefecture noted the anxiety of those who were still left at TSH as a major mental health issue.

The media drew the attention of the public to the vulnerability of elderly people, arguing that they were weak and needed special care and help. The media held that elderly residents of TSH and PRH were a model of Japan's ageing society. Elderly people were the major focus of earthquake related media coverage in late 1998, but by the end of 1999 the focus had shifted to other issues, such as struggling middle-aged survivors and children orphaned by the earthquake, as well as how to prepare for future disasters and emergencies. On the fifth anniversary of the earthquake, the media became more positive and shifted away from looking critically at the past.

Public policy grouped elderly people together with the weak, the disabled and children, and did not take into account gender variations. In contrast, the media reported that elderly women were coping well and were good at making friends, but elderly men were having more problems living alone. TV reports focused more on elderly women than elderly men; both unhappy, lonely elderly women and elderly women with more positive attitudes were featured. TV tended to present women's experiences positively and men's experiences negatively.

The shift in the focus and tone of media coverage over time was parallel with data collected in my fieldwork. The media was critical of the public administration in many ways, if not all the time. It pointed to instances where the administration was not doing enough and emphasized the problems occurring. The public administration was sensitive to the media's reports. The media was one way to give voice to an issue, although it had its own bias and tended to be sensational. The views channeled through the media on an issue did not necessarily coincide with other views, such as those obtained through official public surveys.

On occasion, the media can help to correct the misunderstandings of government about elderly people. It can also be used by the government to inform the public about new policies. This research found that while government surveys tended to reflect traditional thinking, the media drew on more progressive ideas like gender when interviewing those who worked in the field about how elderly men in particular had many problems. Despite their role in promoting sensitivity to gender, the media was biased against seeking information on the financial situation of individual survivors. The media also publicized new initiatives and activities like new categories of staff and staff expansion as part of the reconstruction effort. Some job categories, like the LSA, existed before the earthquake but became better known after the recovery process.

### *Loneliness*

In TV reports, elderly residents of both TSH and PRH often said, 'I am lonely'. Some residents of PRH missed their friends and the sense of community they had at TSH (some of these people may not have experienced such community had it not been for the earthquake). Age was not the only factor affecting the situation of elderly PRH residents; housing type and the extent of community development were also factors affecting resident experience.

From the perspective of critical gerontology, this research questioned the traditional idea that elderly people cannot change and that housing relocation is bad for them. For example, when the media reported that elderly residents were lonely, it blamed the public administration for not treating them well. But this research found that although being alone and being lonely are not the same thing, many elderly residents of PRH, especially men, felt lonely because they could not make friends.

### *Kodokushi*

*Kodokushi* (dying alone) was generally perceived to result from the absolute absence of family. The media used the word to grab attention to their headlines, and reports often featured the number of cases as evidence of a general problem. Although *kodokushi* is an individual problem and it is important to highlight the factors that might put an individual at risk, it is not necessarily the typical experience of elderly PRH residents. The media often reported *kodokushi* cases where the deceased person received care and home visits every day but happened to die when no one was there.

### *Community development*

At the time of research, victims of the disaster were in the process of moving from emergency relief in TSH to normal life in PRH. Community development initiatives like activities, facilities and care services that were effective in TSH were proving difficult to transfer to a PRH setting. This might have been a function of time and no firm conclusions can be drawn at this early stage.

*Gyōsei* (public administration) believed that the Natsu Aki PRH and the Fuyuyama PRH had strong communities because they had skilled leadership that collaborated well with public officials. This research questioned the traditional idea that elderly people living alone are the most vulnerable group in society who need the care and support of their families. The slow community development in PRH could have been due to the lack of volunteers or the fact that they were not used in the same way as they had been in TSH. In time, elderly residents and staff could develop communities by their own initiative.

Coordination between the *gyōsei* and local communities is considered to be the key to the development of a successful community care system. In Japan, the *minsei iin* (welfare commissioner) is a key volunteer person who has established trust and respect in the community.

The establishment of the *jichikai* (self-governing body) was seen as important for community development. The *jichikai* is the official representative of the housing community. Their collaboration with local government enhances health and welfare work. It is not easy for *gyōsei* to operate in a community without a *jichikai*. In PRH, it had been difficult to form a *jichikai* because of the size of the community, its location in high-rise apartment buildings and its development over a short period of time. Some also said it was also due to the age of the residents and the large number of former TSH *yakuins* (community committee members) who had different ideas of how to do things.

No one model will fit all situations. The leader at the Natsu Aki PRH was a *minsei iin*. The leader at the Furuyama PRH was the chief of the *jichikai*. LSAs (Life Support Advisors) also came to play an increasingly important leadership role, but were not able to effectively replace the work done by volunteers at TSH. At places like the Natsu Aki PRH, where other people were more active in taking a leadership role, it might have been difficult for LSAs to play a critical role. But in places like the Fuyuyama PRH, the role for LSAs had more potential. In either place, the LSAs alone could not do all the community work, but could be an important resource for residents and the community.

Japan is a highly health conscious society and health activities are often seen as a means for community development. Some elderly residents said the only people they talked to were doctors and nurses. Health programs are reciprocal; they are effective in getting the local community to work together.

### *Gender and community*

At the community level, management was primarily undertaken by women, whereas formal community politics was undertaken primarily by men (Moser, 1993). This research questioned the traditional idea that men dominate women in this area. Mrs Sakura was taking both a management and a political role. Women were critical for bottom-up policy development and helped policy to directly reflect needs at the local level. Community development was more likely to be successful when a women's group took a major role. The Natsu Aki PRH was active compared to other PRH, and this can be attributed in part to the presence and active participation of Mrs Sakura, the female leader who performed both traditional male roles and roles in women's groups. The changing gender roles in Japanese society were demonstrated in chapter six when comparing the leadership of Mrs Sakura from Natsu Aki PRH and Mr Tanaka, the *jichikaicho* (chief of the self-governing body) at Fuyuyama PRH.

### *Housing types and size*

Housing type was one of the determining factors in the development of community and social networks. Housing environments can change the daily habits and activities of elderly people. Despite the modern housing, the residents of PRH reported problems of isolation and loneliness.

High-rise apartment buildings were not good for promoting community among elderly residents. It was difficult to have human contact with neighbors and to make friends in high-rise PRH. The general process of urbanization, of which the post-earthquake reconstruction is a part, removes elderly people who live alone from their community lifestyles and situates them in more isolated urban environments. Elderly people who lived alone in downtown low-rise houses were less likely to experience loneliness just because of the design of their housing.

Community development in PRH was more difficult because of the large numbers of strangers who had newly moved in. When several high-rise apartment buildings formed a single development, the number of households was between a few hundred up to more than one thousand. The overall size of a housing complex matters for the dynamics of community development (Power and Anesaki, 1999). Community development was seen as particularly difficult in mammoth housing schemes. It is not easy for residents to development relationships with the many strangers that were their neighbors. It was also difficult for LSAs or the *minsei iin* to care for all the residents of a mammoth housing scheme.

Large-scale high-rise modern apartment compounds were created to house people who had lost their homes in the earthquake. This type of architecture might reflect an over-ambitious, over-paternal system of top-down government. The lessons learnt in post-war Europe after the governments provided large scale housing (Power, 1999) were not taken into account. PRH was not necessarily built for elderly people, although some buildings had Silver Housing with barrier-free design on ground floor apartments. Like in older public housing, PRH might be accommodated by younger people as elderly residents die.

This research showed that while high-rise PRH technology improved safety, it did not necessarily produce happiness. Greater and more diverse social networks and better town planning will be required to promote happiness and reduce loneliness.

In densely populated Asian cities, high-rise apartment buildings are unavoidable, making common space, green areas and life style facilities especially important. This is well known, particularly in the Western experience (Gilloran, 1968, Krieger and Higgins, 2002). Gilloran (1968) wrote of the elderly experience of high-rise accommodation,

> Up in the air with their only lifeline the lift, when it works, and a good view of the sky, many of them are tormented by the fear and often the experience of loneliness, which is accentuated should the individual be incapacitated in any way (1968: 117–118).

Gilloran argued that it is necessary to provide common activity spaces and foster good neighborhood relations among residents of high-rise flats. Some attempts were made to factor similar ideas into the development of PRH.

In the Western context, ideas for housing and town planning are ideally derived from the bottom-up, but the massive scale of high-rise apartments requires a top-down approach (Power, 1999). However, in East Asia were high-rise developments are more common, they may not have the same emotional impact or produce the same lack of community. This is understudied in the Asian context (Forrest et al., 2002).

This thesis documented the problems faced by elderly people caused by poor planning, but some of these problems might be resolved over time. This is an area that needs further research.

## Future research

During the post-earthquake reconstruction, Kobe City showed outstanding local government initiatives to construct life care communities in cooperation with local communities. These initiatives should be examined more closely in future research.

Research should also examine how the *jichikai* system, which worked in Japanese traditional society for small-scale communities, can be made to work for large-scale aging communities. This includes the question of how the *rengō* (union) of *jichikai* facilitates community development across various local sites, both active and inactive.

The dynamics of community development in TSH and PRH varied among different sites, variations that might be shared with other communities in the Japanese context. Another country, with a different social, political and cultural context might have a different experience, though there may be some features in common. For example, comparative studies of community development for the elderly population of China might give new insights and interesting findings. This is an important issue in China, as reported in the *China Daily*, November 2004. Future research projects in other ageing societies need to look at the following roles: public and private sector responsibility for aged care; national, prefectural, and local responsibility for aged care; and community, family and individual responsibility for aged care.

The earthquake exposed class divisions in Japanese society. The polarization of communities along class lines should be looked at in future research. Different class groups might need different kinds of support from the public and private sector work, or a different balance between the two. If different class groups are located within the same administrative district, the more affluent area could help to support the development of the poorer neighborhood, or the neighborhoods could be segregated. Because Japan is a racially homogenous society, segregation is far less likely than in a multi-ethnic society.

Future research might also look at the psychological implications of social security and the pension system, especially for low income older people. It was difficult to collect data on this issue for reasons outlined in chapter three.

If an elderly person receives welfare, it does not necessarily mean that they did not contribute to society in their working lives. Many did not work in the formal sector, so they were not entitled to receive pensions. These people were often women who had not worked in the formal sectors, or had not worked long enough and with sufficient continuity to be entitled to pensions.

The psychological and mental health impacts of population ageing in various social and cultural settings should also be looked at in future research. These impacts include different types of legislation and social security systems that affect family structures and gender roles.

Future research could also look at isolation, loneliness and mental illness among poor people. If a person is poor and cannot afford to offer a cup of tea, then they cannot show hospitality. It is difficult for poor people to establish social networks, and elderly people who are poor often become isolated from their children.

Non-Government Organizations (NGOs) assisted in the reconstruction process after the 1985 earthquake in Mexico. The bottom-up approach of NGOs was successful whereas public schools and houses were destroyed and public sector corruption was a serious problem. There was some evidence of NGO activity in Kobe, but further research needs to be done, particularly looking at the process of collaboration between private and public sectors.

When reports began to circulate about the 2008 Wenchuan earthquake in China, volunteers brought water and food, helped the injured and contributed labor and resources to rebuild houses. Mass volunteer efforts had not occurred in China before as disasters were usually responded to by the government. In this case, the government permitted media reporting because of the international criticism it received over its media policy towards the Tibetan unrest in the lead up to the 2008 August Beijing Olympics. The response to the 2008 earthquake can be contrasted with that of the 1976 Tanshan earthquake in China where the media were banned. It is possible that the disaster response triggered a change in the social and political system. Further research on volunteerism in China is needed.

## Contributions to knowledge

This research contributes to knowledge through its original discourse analysis of media data. Media influences social policy in Japan; media reports on loneliness and *kodokushi* were important for nurturing the public response. Despite this, the importance of the media has been underestimated in social policy studies.

The secondary analysis of Japanese public health surveys was also original, especially in that it brought gender into focus, a factor largely lacking in Japanese public surveys. Also, this research was the first to apply the computer assisted qualitative analysis software, QSR NVivo, to a Japanese research context.

My interpretation of media discourse has contributed to current knowledge in an original way. This analysis focused on new earthquake vocabularies such as *kodokushi* (isolated death), *renkei* (coordinated collaboration), and *fureai* (making friends). The key factors in community development among displaced elderly people were leadership, gender roles, the mobilization of existing resources, working together and public health interventions. Measures of success were analyzed to identify a general shift from hard to soft and from physical to emotional success. *Kodokushi* was found to reflect a fear of total lack of family and cases were mainly identified among elderly people living alone. This thesis questioned traditional thinking that sees elderly people who live alone as 1) the weakest and most vulnerable group in society who need support from younger generations, 2) unable to change or adapt to new housing environments and 3) dominated by men in community politics.

This research concluded that people who had formerly been involved in well organized traditional communities were displaced by the earthquake into new housing with few resources to build new communities, despite having a more modern living environment. It argued that this is a more general development problem for Japanese society. The research studied the case of Mrs Sakura, whose innovations were successful in adapting traditional *minsei iin* systems to new community initiatives in the context of large scale and densely populated developments with many elderly residents.

This research also differed from other studies of earthquake reconstruction by looking at the longer term rather than immediate effects. This research looked at the recovery process five years after the disaster happened.

## Conclusion

This research looked at the recovery process following a natural disaster. These processes are likely to be present in similar disasters but generalizations about size or extent are problematic. The complexity of large scale natural disasters means that each case will be unique, though with many common characteristics.

The 1995 Great Hanshin Earthquake revealed the experiences of unsupported and low-income elderly people. Before the earthquake, they did not feature in the media or in public policy. The earthquake affected these people the most and their concentration in camps made them highly visible.

Elderly people typically fear living alone (Hashimoto, 1996). Living alone is generally not a problem for people who are healthy, but if something happens to a person's health living alone can be very difficult. Elderly people living alone often rely on family or some other service to respond to an emergency.

The research found that *renkei* (coordinated collaboration) between *gyōsei* (public administration) and local groups of volunteers is important for developing community in large high-rise apartment buildings. Through *renkei*, a community can better identify the resources available and how to access and utilize them.

At the time of research, some inroads had been made towards developing community in PRH, but more was needed by *gyōsei,* local groups and other participants in civil society.

It was difficult to develop community in high-rise PRH, but it is not clear if this was because of architecture, the short time scale or the lack of social work. The problems identified in this research may improve over time, but the architecture will not change and social policy might not either. Because there is limited space in cities, there are few options other than to build more high-rise apartment buildings in the future. It was not possible to rebuild the low-rise dwellings that were destroyed in the earthquake.

The public housing provided was not well thought through (Hirayama, 2000). The community interventions observed during this research revealed that elderly residents living alone could be active in developing their own communities while those who lived with their families tended to stay at home. Elderly women living alone were typically more active in their communities than elderly men and it may therefore be that elderly men living alone are a more at-risk group. Regardless of gender, living alone is not a problem in itself; a person can live a long and independent life if adequately supported.

Like many other societies, Japan is going down a demographic road. In future, ideologies of the family will have to provide much more non-family support, will have to do so in different ways from the traditional family support system, and will need to think about future housing provision in terms of encouraging community action among isolated older people. New ways of producing family care and social cohesion that do not rely on the exploitation of women are urgently needed (Wilson, 2000: 168).

Disasters have long and lasting affects for a society's most vulnerable people. In the case of this research, these were elderly people without money or families. Official statistics and the media make their own interpretations of what is going on and the workers on the ground reproduce many of their views and some prejudices of their own.

The loneliness of elderly people is increasingly visible to the public. The second UN World Assembly on Ageing took place in April 2002 in Madrid, twenty years after the First World Assembly on Ageing was held in Vienna in 1982. The decisions of Madrid focused on loneliness, isolation, neglect, and elderly abuse. Future research should look into these issues.

The 2008 Wenchuan earthquake in Sichuan, China, exposed the difficulties faced by elderly people left alone in mountain villages damaged by the earthquake, although no official figures are available (Chan, 2008). Chinese society is experiencing rapid population ageing due to its population policy and the migration of younger generations to urban centers seeking economic opportunities. Elderly people are left alone in rural areas with no social security. Although the immediate response to the earthquake has ended, more research on the long term recovery is needed. The findings of the Kobe research may contribute to identifying the groups most vulnerable to the long-term effects of the disaster in Sichuan.

# *Epilogue*
# *My personal experience of the Great Hanshin Earthquake*

I learnt of the earthquake on 17 January 1995 from Bangkok, where I'd just returned from visiting Myanmar, Northern Thailand, to participate in a mission responding to the HIV/AIDS epidemic. From my Bangkok hotel room I was able to watch the untranslated NHK Kobe emergency broadcast. NHK Kobe was the first news provider to report from the scene of the disaster.

All the Japanese on the street, even strangers, were talking to each other: 'Are you Japanese? The death toll is 2,000. The updated number is 3,000; no, 5,000. It was a big disaster; all the houses were destroyed.' I rang my home from Bangkok but it was impossible to connect because many of the phone lines were destroyed.

The next day I flew to Narita International Airport outside Tokyo and took the first bullet train to Osaka that operated after the earthquake. Going east from Osaka, the bullet train resumed service after three days but it was several months before the westbound line opened.

When I arrived home I found that all my family were safe but the inside of the house was a mess. I said to my mother, 'It has been three days. Why haven't you picked up these very important things from the floor?' She answered, 'I have been working hard to clean up but have not been able finish the job. First I cleaned up the dangerous things like the broken glass. I had to throw away all these East European wine glasses and British made China tea cup sets.' I stayed at home for three days before returning to work in Tokyo. Although I worked hard, I was only able to clean up my room, which was mostly a mess of books. I did not have time to work on other parts of the house.

During this time, aftershocks were still occurring. The aftershocks were of greater magnitude than usual earthquakes. Having grown up in Japan, I have experienced many minor earthquakes and have never felt afraid of them. I rather enjoyed the 'shaky moody' earth when I was small; I thought it was like riding on something at an amusement park. At school, I obediently followed the evacuation instructions but did not consider them a serious matter. After experiencing these aftershocks, I started to fear minor earthquakes. They might have been of lesser magnitude than the first earthquake, but they were potentially more dangerous because they could destroy already damaged buildings.

By the time I had returned to my work in Tokyo, the severity of the disaster was widely known. Despite this, it did not feature prominently and on a daily basis in the Tokyo news. A colleague of mine said, 'Is it really happening in this very Japan? Being away in Tokyo, we do not feel it.' This comment saddened me. The Aum terrorist attack that occurred in the Tokyo subway in March drew more public attention in Tokyo from the recovery in Kobe and Osaka.

I visited my family in Osaka a few more times over the next few months. By March, the house had been tidied and my mother took me to see the most affected area. She said, 'We should not just go and look. There are few toilets and visitors need a toilet. We should go as volunteers.' As the Hankyu train line that passes through the middle-class mountain side area had not then been reopened, we took the Hanshin line through the downtown working class area. The train was crowded and many of the passengers were dressed in working clothing and walking shoes. This was how people dressed after the earthquake, not in business suits. The journey took much longer than usual as the train could not travel as fast as usual because some parts of the line were still not repaired fully.

Looking out of the window of the train I was shocked to see the extent of the city's destruction. Many houses were burnt down by the fire triggered by the earthquake. The destruction was overwhelming and I could not help my tears. In a way I felt like I had entered a time machine and slipped back 55 years, when the whole city was destroyed by heavy bombing in the war. I was not yet born at the time of the war and I had only known affluent post-war Japan, but I remembered a black and white photo of my grandfather sitting in Hiroshima after the atomic bomb. A young professor at Osaka University whose brother-in-law had been killed by the atomic bomb in Nagasaki, he had travelled with a colleague to Hiroshima, saying, 'We must see this historic event with our own eyes.'

My grandfather's brother-in-law was a second year student of Nagasaki Medical College. He had wanted to study law at Tokyo University, following the path of his father, but his mother begged him not to go to Tokyo, assuming that Tokyo would be bombed and that a degree in law would lead him to serve in the military. Sadly, this choice ultimately led to his death by atomic bomb two years later. My grandfather was determined to contribute his expertise in science and technology to the development of Japan and to devote his life to the peace, wellbeing and prosperity of the world. Years later he developed the world's largest electro-microscope which only Japan and France now have.

I left Japan a few months later to take up a new job at the World Bank in Washington DC. This job allowed me to return home five times a year on the way to other Asian countries. When I returned home again, my parents' house was fully repaired and renovated at the cost of JPY 8 million. Although some home owners received government subsidies from the disaster-relief scheme, a city inspector decided that the damage to my parents' house was not as serious as in other areas so they did not receive any support.

A granduncle of mine in Kobe was one of the elderly people who succumbed to dementia during the post-earthquake recovery. Because his house was destroyed, he stayed in a hospital as many other elderly people with health problems did. Hospitals were seen to be a safer place for elderly people and generous medical insurance encouraged many elderly people to stay in them, a phenomenon called social hospitalization. In his youth, he was a university professor and a very bright intellectual, but he became senile shortly after being hospitalized. He passed away in the hospital after a year.

After working in the World Bank Asia Department for a few years, I had an opportunity to return to graduate school with the World Bank Scholarship. I was interested in elderly people for three reasons. First, population ageing was emerging as an issue in China, the country where I had done most of my work, and in other countries both developed and developing. Second, the challenge of population ageing was featured daily in Japanese newspapers. By studying ageing, I could do work in Japan for international organizations. Third, my parents were getting older. I am their first child and I felt guilt for leaving them to work abroad. Also, my grandmother and I are very close; she is my best friend.

While living in London, I returned to Japan for Christmas and New Year's holidays. There I saw the media coverage of the elderly residents of post-earthquake housing communities. Many compared their situation to other disasters they had experienced in their lifetime, such as the 1918 Kobe flood and the Second World War. I found this personally interesting and my supervisor encouraged me to pursue it as a research project.

It was difficult to gain access to temporary housing sites in Kobe, but I managed thanks to Ms. Kiyoko Okawa, public health nurse volunteer for JOCS who had worked hard to gain the trust of the residents. Her manner of interacting with the residents reminded me of Prof. Mieko Kamiya, a writer, psychiatrist and philanthropic doctor for leprosy patients who supervised my mother at Kobe College. By looking at Ms Okawa, I recalled Prof. Kamiya's poem, 'To a leprosy patient', published in 1966. I wish to conclude this book with my translation of this poem:

*To a leper*

Vacant eyes losing light
Body losing limbs
Dropped its body on the medical examination bed
Dear Leper
I lower my head in front of you.

You remain silent
Even smiling slightly
Oh, however, the silence and the smile was won
After a long journey of struggle.

You are living at the edge of fate
Even trying to escape in vain
Caught by the iron hand in day and night
You have lived for ten years, twenty years.

Why you and not us?
You have vicariously suffered on our behalf
You have been deprived of everything as a human being
You have suffered and outran the torments of hell.

Please forgive us, you who suffer from leprosy!
We who float shallowly and lightly on the sea of life
Manipulating pointlessly nice sounding words like God and soul.

Crying out to the core and lowering my head
You remain silent
And even keeping smile
In your face that was painfully distorted.

# *Appendixes*

## Appendix 3-1: Fieldnotes list

Table A3-1.1: Observation

| Place | Date | Time | Length of report (words) |
|---|---|---|---|
| Haruyama TSH | 3 April 1998 | 8:00–17:00 | 2062 |
| | 3 Sep 1998 | 8:00–17:00 | 2400 |
| | 18 Nov 1998 | 10:00–17:00 | 3264 |
| | 25 Nov 1998 | 10:00–17:00 | 1708 |
| | 2 Dec 1998 | 10:00–17:00 | 1007 |
| | 9 Dec 1998 | 10:00–17:00 | 1400 |
| | 16 Dec 1998 | 10:00–17:00 | 1059 |
| | 23 Dec 1998 | 10:00–19:00 | 798 |
| | 30 Dec 1998 | 10:00–17:00 | 596 |
| | 6 Jan 1999 | 10:00–17:00 | 1043 |
| | 13 Jan 1999 | 10:00–17:00 | 432 |
| | 27 Jan 1999 | 10:00–17:00 | 954 |
| | 24 Feb 1999 | 10:00–17:00 | 2750 |
| Seishin 7th TSH | 4 Sep 1998 | 9:00–17:00 | 1700 |
| Natsu Aki PRH | 4 Feb 1999 | 11:00–13:40 | 6863 |
| | 5 Feb 1999 | 11:30–13:40 | 3633 |
| | 15 Feb 1999 | 10:40–14:10 | 4544 |
| | 16 Feb 1999 | 10:00–13:00 | 6944 |
| | 25 Feb 1999 | 10:00–11:20 | 2976 |
| | 8 July 1999 | 13:00–18:00 | 3568 |
| | 9 July 1999 | 10:00–12:20 | 6364 |
| | 17 July 1999 | 10:30–14:30 | 3756 |
| Fuyuyama PRH | 9 Feb 1999 | 9:30–11:20 | 4598 |
| | 19 Feb 1999 | 9:00–12:10 | 6097 |
| | 25 Feb 1999 | 12:00–13:20 | 3208 |
| Kobe SH PRH | 1 June 1999 | 15:00–20:00 | 2807 |
| Nada PRH | 10 June 1999 | 10:00–16:00 | 1905 |

**Typical field research diary days**

*A Day of tea party at Haruyama TSH*

| | |
|---|---|
| 7:00 a.m. | Leave home |
| 8:20 a.m. | Kobe Sannomiya Station |
| 9:00 a.m. | Seishin Minami Kobe Subway Station |
| 10:00 a.m. | Arrive at Haruyama TSH. |
| | Open and clean the Haruyama TSH Community Center and start preparing for the tea shop |
| 12:00 | Lunch |
| 1:00 p.m. | Serving at a tea party as a volunteer waitress |
| 3:00 p.m. | Tea shop ends |
| | Cleaning the kitchen and the community center |
| 4:00 p.m. | Staff meeting |
| 5:00 p.m. | Lock the community center and return home. |

*A day of home visit accompanying the volunteer public health nurse at Haruyama TSH*

| | |
|---|---|
| 7:30 a.m. | Leave home |
| 8:45 a.m. | Meet up with the public health nurse at the Hankyu Umeda (Osaka) Station |
| | Travel to Kobe together and talk with her on the train. |
| 10:00 a.m. | Arrive at the Haruayama TSH. |
| | Start making a home visit to each household still left |
| 12:00 | Lunch and continue a home visit |
| 4:00 p.m. | Leave the Haruyama TSH and return home |

*A day of a tea party and health promotion and counselling program at the Natsu Aki PRH*

| | |
|---|---|
| 7:50 a.m. | Leave home |
| 10:00 a.m. | Arrived at the assembly hall of the Natsu Aki PRH. |
| | Sit at the tea party and observe |
| 12:00 | Interview key informants |
| 1:00 p.m. | Leave and return home |

*A day of lunch service at the Natsu Aki PRH*

| | |
|---|---|
| 8:00 a.m. | Leave home |
| 10:20 a.m. | Arrive at the subway station and wander around the area. |
| 11:00 a.m. | Arrive at the assembly hall of the PRH and look around. |
| | Take notes from the notice board. |

11:30 a.m. People arrive
1:00 p.m. Helped cleaning the dishes and room
1:45 p.m. Leave

*A day of mini-day care centre at the Natsu Aki PRH*
8:20 a.m. Leave home
10:40 a.m. Arrive at the assembly hall of PRH
11:00 a.m. Mini day care service started
Physical exercise
11:30 a.m. Lunch
12:30 Game
1:30 p.m. Afternoon tea
2:00 p.m. Sing a song
3:00 p.m. the program ends
Clean
3:30 p.m. Leave

*A day of one of the first* Yūai *home visit activities after a monthly birthday party at the Natsu Aki PRH*
1:30–3:00 p.m. Monthly birthday party at the assembly hall
3:30 p.m. Staff Meeting at Tatami room
4:00 p.m. Yūai (Friendship and Love) home visit
5:15 p.m. Leave for home

*A Day of breakfast program at the Fuyuyama PRH*
7:15 a.m. Leave home
9:00 a.m. Arrived at the Fuyuyama station. Walk around and take some photos
9:20 a.m. Volunteers at the Catholic church start a silent prayer
9:30 a.m. The breakfast service starts
11:10 a.m. Interview with a key informant
12:30 Leave

*A day of Health promotion program at the Fuyuyama PRH*
12:00 Arrive at the assembly hall of the PRH
12:15 p.m. Observation at the health promotion program
2:00 p.m. The program ends. Interview

Table A3-1.2: Interview notes list

| | Where | Who present | Date, How long | Length of report (words) |
|---|---|---|---|---|
| *Minsei Iin* (Welfare commissioners) of Natsu Aki PRH | Assembly Hall of the Natsuyama PRH | Mr Takada and Mrs Sakura | 4 Feb 1999 11:00 to 12:00 (1 hour) | 2976 |
| Chief of the self-governing body of Akiyama PRH5 | Assembly Hall of the Akiyama PRH | Mr Kamiyama, the Chief | 16 Feb 1999 11:45 am to 13:00 (1 hour and 15 min.) | 313 |
| Chief of the self-governing body of Fuyuyama PRH | Assembly Hall of the Fuyuyama PRH | Mr Tanaka, the Chief, and two LSAs | 9 Feb 1999 9:45 am to 11:10 am. (1 hour and 25 min.) | 4598 |
| LSA at the Fuyuyama PRH | LSA office of the Fuyuyama PRH | Ms Suwa, LSA | 10 Feb 1999 11:10 am to 12:10 (1 hour) | 1897 |
| Director of the old people's home in Ashiya | The old people's home | Mrs Ichikawa, Director | 22 July 1999 10:00 –14:30 (4 hours and 30 min.) | 7749 |
| Director of the Department of Health and Welfare of Kobe City | Office of the Director | Director and three staff of his office (Chief of Health, General Affairs, and Accounting and Administration) | 26 Jan 1999 11:00–12:00 (1 hour) | 2877 |
| NHK Kobe TV reporter | Italian restaurant near the NHK Kobe | Ms Aya | 31 May 1999 12:30–14:15 (1 hour and 45 min.) | 1913 |
| *Kobe Shinbu*n newspaper journalist | *Kobe Shinbu* Office | Ms Isobe | 10 June 1999 17:00–18:00 (1 hour) | 1780 |

# Appendix 3-2: TV Coverage 1999

Table A3-2.1: TV coverage 1999 list

1. 1998/11/24/18:35 *Channel 10 News.*
2. 1998/12/05/18:10–18:40 NHK 'Searching for housings which we can support each other: Choice of elderly people at the time when soon will be the fourth year after the Earthquake' (Recorded in video tape).
3. 1998/12/17/23:15–23:45 NHK Educational *Home Economics of Men and Women,* 'Human beings and Living,' Lectured by Akiko Nakajima, Mejiro University College.
4. 1998/01/08 Fri/21:30–22:00 Sun TV *EYE f forte* 'Three years in temporary shelters and a New Year at a public reconstruction housing.'
5. 1999/01/10 Sun/18:30–19:00 Sun TV *Phoenix Hyogo: Disaster Reconstruction* 'Towards the creative post-Earthquake Reconstruction.'
6. 1999/01/10 Sun/23:15–23:40 NHK *Discovering New Japan* 'Make a with on Nabe (cooking pot)' Kobe Nagata 4th year gathered around the nabe used at a evacuation centre at the time of the Earthquake.
7. 1999/01/10 Sun/24:15–24:45 TV Yomiuri *NNN Documentary '99* 'Four years after the earthquake.'
8. 1999/01/11 Mon/17:40– Sun TV *First news in the evening* 'Earthquake Series (1) 'The fault line will tell the memory of 1.17: Kita-Awa village,' Broadcasting from the Memorial Park.
9. 1999/01/11 Mon/17:54– Kansai-TV Super News Kansai 18:30– Earthquake year 4 Series (1) Expiration date for leave is March—The fact Still 5,800 households are living at a temporary shelter.
10. 1999/01/11 Mon/16:05– ABC-TV *Wide ABC-DE-su* 18:00– 'Four years after the Earthquake.'
11. 1999/01/11 Mon/18:00 NHK *Newspark Kansai* 'Four year after the earthquake, the last temporary shelters.'
12. 1999/01/11 Mon/18:30 Mainichi-TV *Now* 'Remember 1.17—Has the economy of Kobe recovered?'
13. 1999/01/11 Mon/18:23– Yomiuri-TV Scramble 'Living at a temporary shelter (1) Testimony to the fifth year,' 18:30–18:32 Current situation report of the regular check points 'Report of the current situation,' 18:35– 'Who created temporary shelter residents with what kind of thoughts.'
14. 1999/01/11 Mon/21:30– Sun TV *Eye f Forte* 'Cooking class to cheer up elderly people living in a temporary shelter.'
15. 1999/01/12 Tue/11:30–noon *Square* 'Filmed by a housewife.'
16. 1999/01/12 Tue/17:40– Sun-TV *First news in the evening* 'Earthquake series (2) What after the break-down of the special temporary shelter with special care system for elderly people and disabled people.'
17. 1999/01/12 Tue/17:54– Kansai-TV *SuperNews Kansai* 18:30– 'What was left behind by the reconstruction plan with a priority on housing.... Appeal from a shop owner who had to close his business.'
18. 1999/01/12 Tue/16:05– ABC-TV *Wide ABC DE-su* 18:00–
19. 1999/01/12 Tue/18:00 NHK *NewsPark Kansai* 'Recording the Earthquake.'
20. 1999/01/12 Tue/18:30 Mainichi-TV *Now* 'Want to make a new factory.... The dream and the reality of a factory owner who risked with a special plan, 'Earthquake Special Exception'.

21. 1999/01/12 Tue/18:23– Yomiuri-TV *Scramble* 'Temporary living' (2). 'Kill me in the third winter father begged his son.'
22. 1999/01/12 Tue/21:30– Sun TV *Eye f Forte* 'The end of this March the temporary shelter will expire.'
23. 1999/01/12 Tue/23:00– NHK *News 11* 'Four year after the earthquake—Hospital information network system started to function.'
24. 1999/01/12 Tue/23:35–23:45 NHK *Reading Tomorrow* 'Fourth winter of Kobe.'
25. 1999/01/13 Thu/17:40– Sun TV *First news in the evening* 'Earthquake Series (3) Disagreement with public administration—Feelings inexpressable by the residents.'
26. 1999/01/13 Wed/17:54– Kansai-TV *SuperNews Kansai* 18:37–18:47 (Recorded Video Vol.1–11) 'Life Support Advisor (LSA) visits around (patrol) Barrier-free housing—An ideal type housing with care that disaster-affected area try to realise through the experience.'
27. 1999/01/13 Wed /16:05– ABC-TV *Wide ABC DE-su* 18:00–.
28. 1999/01/13 Wed /18:00 NHK *NewsPark Kansai* 'Four years of those who lost their house in the earthquake and now living outside the prefecture (recorded Video vol.1–9).'
29. 1999/01/13 Wed /18:30 Mainichi-TV *Now* 'Group Home, a housing good for aged people—suggestions from experiences at temporary shelters (recorded Vol.1–12 last 6min. only).' (Rebroadcasting of Tsukushi Tetsuya News 23, Recorded vol.2–5)
30. 1999/01/13 Wed /18:23– Yomiuri-TV *Scramble* 'Wanting to see your smile and to show my smile: A small happiness a girl found in Kobe.'
31. 1999/01/13 Wed /21:30– Sun TV *Eye f Forte*
32. 1999/01/13 Wed/22:00– ABC-TV News Station 22:55– 23:10 '3 day series of earthquake: closed space—the final housing of the earthquake evacuees' (recorded Video vol.1–13).
33. 1999/01/13 Wed /23:15 Mainichi-TV *Tetsuya Chikushi NEWS 23* (recorded vol.2–5 10 min.)
34. 1999/01/14 Thu/11:45–11:55 *Square* 'Four Years after the disaster: What for do they lead their lives.'
35. 1999/01/14 Thu/17:40– Sun TV *First Year in the Evening* 'Earthquake Series (4) Struggle for Sugahara Market Reconstruction, Reported by Journalist Kiyoshi Kuroda.'
36. 1999/01/14 Thu/17:54– *SuperNews Kansai* 18:30– ' 'No job' Economic depression prevents post-earthquake reconstruction—Report from a temporary factory: Chemical Shoes Factory.'
37. 1999/01/14 Thu/16:05– ABC-TV *Wide ABCDE-su* 18:00
38. 1999/01/14 Thu/18:00 NHK *NewsPark Kansai* 'The situation of earthquake area volunteers.'
39. 1999/01/14 Thu/18:30 Mainichi TV *Now* 'Wants to maintain the relationships—Life reconstruction of old people and collective housings.'
40. 1999/01/14 Thu/18:23– Yomiuri TV *Scramble* 'Temporary Living (1) Please come again Although moved to a permanent housing, I am aged and alone. Old people and my home' (recorded Video Vol.2–2).
41. 1999/01/14 Thu/19:30–20:00 NHK Educational *Future Living Together* 'Earthquake changed my shoe making.' (Recorded Video Vols.2—3 30 min)
42. 1999/01/14 Thu/21:30– Sun TV *Eye f Forte* 'Scenery of towns in the Hyogo Prefecture.'
43. 1999/01/14 Thu/21:30–22:00 NHK *Close Up Modern Society* 'Isolated old people—Report from a public reconstruction housing after the earthquake' Rebroadxasting of the 5 Dec programme 4 years after the earthquake' (Recorded Video Vol.3—3 30min.).
44. 1999/01/14 Thu/23:00–23:20 NHK *Hello Japan We live on the Earth* 'Remember the earthquake' (recorded video vol.3–4 20min.)
45. 1999/01/14 Thu/20:54– Mainichi TV *Tetsuya Chikushi* NEWS 23 23:20– 'Care the aged—suggestions from a temporary shelter'(Rebroadcasting the Program MBS 1999/01/13 18:30).

46. 1999/01/15 Fri/18:10–45 NHK *Holiday Japan.* 'I know it is hard but let's go for it. The Hanshin Earthquake Reconstruction Support Factory' (Recorded Video Vol.2–6 45min.)
47. 1999/01/15 Fri/17:54– *SuperNews Kansai* 18:30 'The feelings the citizens convey on January 17, the memorial day.' Memorial by people 'High school volunteer visits a temporary shelter.'
48. 1999/01/15 Fri/16:05– ABC TV *Wide ABCDE-su* 'Living at a temporary shelter was much better than at public reconstruction housing.' 18:00 'Why they fall into alcoholism?'
49. 1999/01/15 Fri/22:50– ABC TV *News Station* 'Four years after the earthquake (3) Rainbow to the earthquake orphans: Finally the Rainbow House has been built.'
50. 1999/01/15 Fri/23:10–23:20 'Four years after the earthquake: Hell of the destroyed flat buildings—Report from Kobe (Recorded Video Vol.3–8 10min.)
51. 1999/01/15 Fri/21:30– Sun-TV *Eye f Forte*
52. 1999/01/15 Fri/20:54– Mainichi TV *Tetsuya Chikushi NEWS 23* 23:30– 'Then from the earthquake' reported by Kiyoshi Kuroda.
53. 1999/01/15 Fri/21:12– Yomiuri TV Bunchin Katsura 'Port Island Temporary Shelter, HAT Kobe Reconstruction housing.' (recorded Video Vol.4–1)
54. 1999/01/16 Sat/18:10–40 NHK *Dispatch base '99* 'The situation and difficulties the Earthquake volunteers face in the fourth year.' (Recorded Video Vol.4–2 30min.)
55. 1999/01/17 Sun/01:43—ABC TV *Telemen* 'In the fifth year of the earthquake. Desperation of temporary shelter...people left behind.'
56. 1999/01/17 Sun/02:41–05:30 Sun TV Re-broadcasting of the programs aired in the past one year 'Recorded Video Vol.5–1' 'Kobe Shoes, Bakery factory of the disabled people', (98.7.14) 'Brick Street' (98.2.27) 'Out of the Prefecture evacuees' (98.2.25) 'Wish to return' (98.11.4) 'Leave the temporary shelter ALIVE, not dead, 'The heaviness of Live' (98.11.10) 'Future of the evacuees who now live in eastern Japan ('Tokyo area' (98.11.11) 'Efforts to get the customers back to the KOBE port' (98.8.6) 'Over the Earthquake' (98.11.19) 'Recovery of Mind, theatre 'Love and Circus' (98.6.12, 8.7) 'The first public housing allowed to keep an animal pet in Japan' (98.12.14) 'Poem and essay of the feelings of the Earthquake sufferers' (98.2.12).
57. 1999/01/17 Sat/03:32–05:15) Yomiuri TV Special Report 'Films from the earthquake' (video Vol.4–4).
58. 1999/01/17 Sun/06:00–07:00 Sun-TV 'The road to reconstruction–Film 'Town of human beings People in Takatori'. '
59. 1999/01/17 Sun/10:00–11:00 Mainichi TV *Remembering 1.17* 'With Testimony of 30 people—Has your attitude to life changed after the earthquake?' (Video Vol.5–3 48min.)
60. 1999/01/17 Sun/11:00 NHK 'Four years after the Hanshin Earthquake, Exploring Kobe, Using Reconstruction maps (Video Vol.5–4 50min.)
61. 1999/01/17 Sun/10:30–12:00 NHK-BS 1 'Four years after the earthquake, Search for making a new town—how to make a new living.'
62. 1999/01/17 Sun/18:00 *NewsPark Kansai* 'Special Report—Four years after the earthquake' (Video Vol.5–5).
63. 1999/01/17 Sun/18:00– Mainichi TV Special Report 'Four years after the earthquake—the last winter at a temporary shelter.'
64. 1999/01/17 Sun/19:00– Sun-TV *Requiem and Steps III* 'Each people's four years—letters from a temporary shelter resident—Questions from public reconstruction housings' (Vol.5–6).
65. 1999/01/17 Sun/21:00–21:50 *NHK Special* 'Reconstruction Left Behind' Four years after the earthquake, 9 trillion yen for reconstruction has been spent on what? When the living will come back? (Vol.5–7).

66. 1999/01/17 Sun/22:00–23:00 NHK BS-1 *Sunday Special* 'Photo Journalist Takeji Kumagai.' (Video Vol.6–1)
67. 1999/01/17 Sun/24:15–24:45 Yomiuri TV *Document '99* 'Four years after the earthquake.' (Vol.5–8)
68. 1999/01/17 Sun/24:20–25:20) Mainichi TV *Film90* 'Home Care: Four years after the earthquake—the current situation of care for elderly people' (Vol.6–3).
69. 1999/01/17 Sun/25:25–26:29 Yomiuri TV Remember the Earthquake (3) Evacuation Center—Tears and Good-bye (Rebroadcasting the program of 1995) (Vol.6–3)
70. 1999/01/17 Sun/26:29–1999/01/18 Mon/03:00 Yomiuri TV 'Remember the Earthquake (4) City reconstruction will be to whose hands—election of city major two months after the Earthquake–voting from an evacuation centre—Ashiya City.'
71. 1999/01/18 Mon/17:54– Kansai TV *SuperNews Kansai* 'We wanted to move back to our newly reconstructed market bazzar.... Fifth year of the earthquake—each people's first step.'
72. 1999/01/18 Mon/18:23– Yomiuri TV *Scramble* 'Lost my shop and lost my house. Some day...these four years.... I will .... I am a 60 year old MAN and I will not give up ...I will not be a loser.'
73. 1999/01/18 Mon/16:00– ABC TV *Wide ABCDE-su* 'Disaster Prevention in the USA.... San Francisco.'
74. 1999/01/18 Mon/18:30– Mainichi TV *Now* 'Fifth year after the earthquake.... Charity 10 km walk: what we saw...what was here.'
75. 1999/01/18 Mon/19:30 NHK Educational *The welfare of Japan Islands* 'Does Care Housing Support the Aged?–Tottori Prefecture. (Video Vol.6–4 30min.).
76. 1999/01/18 Mon/22:00– ABC TV *News Station* 'Four years after the earthquake—between public supports.'
77. 1999/01/18 Mon/21:30– Sun TV *EYE f Forte* 'Volunteer activities in the fifth year after the earthquake.' (Vol.6–5)
78. 1999/01/18 Mon/23:35–23:45 NHK *Reading Tomorrow* 'Earthquake and issues of housing reconstruction' (Vol.6–6).

# Appendix 3-3: List of TV Coverage in 1999 and 2000

Table A3-3.1: List of TV Coverage in 1999 and 2000

1. 1999/06/13 7:45-8:00 TV: NHK Kobe 'Reconstruction '99: The current situation of housing in the earthquake-affected area' series. No. 2 Collective House.
2. 2000/01/07/18:35 Ch 10 News Scramble. 'Long-term report on the troubles and events after the earthquake: five years of disaster effect'.
3. 2000/01/07/17:54 Ch. 8 SuperNews Kansai 'Five years since the earthquake'. The key to the life reconstruction is 'relationship' between human beings.
4. 2000/01/07/18:30 Ch. 6 Wide ABCDEs. The reason why they cannot leave the temporary shelter.
5. 2000/01/07/18:30 Ch. 4 MBS Mainichi News Now 'Remember the day. Five years since they left Kobe. Turning point for those disaster-affected now living outside the Prefecture.
6. 2000/01/07/ 21:30-22:00 Sun-TV EYE Two years in public reconstruction housing. Towards life reconstruction.
    a. Reconstruction Symposium. Each becomes independent.
    b. Life reconstruction: Case of Ms. Chieko Ueda. Two years since she moved to public reconstruction housing.
7. 2000/01/09/Sun/07:45-08:00 NHK Ch. 2 Special feature of post-earthquake reconstruction. Towards the fifth anniversary of the earthquake.
8. 2000/01/09/Sun/24:25-2455 Ch.10 Yomiuri TV 'Document '00. Five years of a victim of the Hanshin Earthquake'
9. 2000/01/10/Mon/17:54 Kansai TV Ch.8 Super News Kansai. Five years after the earthquake . The end of the ninety years history of the Sugahara Market. What are those doing who supported the Market?
10. 2000/01/10/Mon/18:23 Yomiuri TV Ch.10 Scramble. Five Years of the Nagata ward Taisho street market.
11. 2000/01/10/Mon/18:00 ABC TV Ch.6 Wide 630 News 'Evacuation with the wheelchair' Series on the fifth anniversary of the earthquake to think about what was left after the earthquake. Community development. To form town with prevention for the disaster with the special attention to the weak. What is the current situation of volunteers? What is the ideal life for old people. Series will continue until 17 of Jan (except over weekend).
12. 2000/01/10/Mon/2:00-22:45 NHK Edu Ch.12. ETV Special Feature. 'Five years after the earthquake. The path of mind since the morning. The day a junior high school student became a grown-up.' This is the first one of series of three. Reports the three generations who live in Kobe City Higashi Nada Ward Motoyama District. In this first report, what was the experience of teenagers?
13. 2000/01/10/Mon/21:30-22:00 Sun-TV Eye news. Disaster prevention learning from the earthquake experience.
14. 2000/01/11/Tue/05:00- NHK. Ch.2 Good morning Japan. 'Fifth year anniversary of the earthquake. Turning point of the disaster-affected markets.'
15. 2000/01/11/Tue/17:54- Kansai TV. Ch.8 Super News Kansai. Five years after the earthquake. We will not forget the children. Memorial parents built for their lost children.
16. 2000/01/11/Tue/18:23- Yomiuri TV. Ch.10 Scramble News. Memories in the memorial. Gifts from the stone JIZO (a guardian deity of children).

17. 2000/01/11/Tue/18:30- ABC TV Ch.6 What the rescue dogs are doing now.
18. 2000/01/11/Tue/19:00- NHK News7 Ch.2 Five years since the earthquake. The issue of the disaster prevention helicopter.
19. 2000/01/11/Tue/21:00- NHK News9 Ch.2 Five years since the earthquake. The safe policy of highways
20. 2000/01/11/Tue/21:30- Sun-TV News EYE f 'Disaster Prevention policy from the learning experience ④ Telephone'
21. 2000/01/11/Tue/22:00-22:45 NHK Edu. Ch.12 ETV Features 'Five years after the Earthquake. The path of mind since the morning. ② Old people at the Temizu Temporary Shelters'. (Collective house, special type for disabled and old people. LSA is Ms. Michiko Kuwahara).
22. 2000/01/11/Tue/23:35-23:45 NHK Ch.2 Reading tomorrow 'Five years since the Great Hanshin Earthquake.
23. 2000/01/12/Wed/05:00 NHK Ch.2 Good morning Japan. Disaster Prevention.
24. 2000/01/12/Wed/17:54 Kansai TV Ch.8 Super News Kansai. The study group of academics proposed to the Kobe City major Sasayama from their study on post-earthquake reconstruction covering infrastructure and so on. 'Ageing society at the public reconstruction housings'. 'Earthquake series. Housing was rebuilt, but…'
25. 2000/01/12/Wed/18:23 Yomiuri TV Ch.10 News Scramble '17 year old who made a vow/ pledge/ oath to the dead father. Five year path of the earthquake orphan.'
26. 2000/01/12/Wed/18:30 ABC TV Ch.6 Wide ABCDEsu 'Housing arrangement to support each other'
27. 2000/01/12/Wed/18:30 Mainichi TV (MBS) Ch.4 Now 'Memorize the day 1.17' ③ Built a high-rise building instead of tenement house. Warmth coming back.
28. 2000/01/12/Wed/19:10-45 NHK Edu Ch.12 Live sparklingly. 'Wish to convey the heart from the earthquake-affected area.'
29. 2000/01/12/Wed/19:00 NHK Ch.2 NHK News 7 'Disaster prevention for highways. Disaster prediction system.'
30. 2000/01/12/Wed/21:00 NHK Ch.2 NHK News 9 'If the company (enterprise)'s disaster prevention policy has developed?'
31. 2000/01/12/Wed/21:30-22:00 NHK Ch.2 Close Up Gendai (modern society) 'Five years since the earthquake. Small and middle-size enterprise (company) standing in difficulties. Searching for reconstruction.'
32. 2000/01/12/Wed/22:00-22:45 NHK Edu Ch.12. ETV Special Feature. 'Five years after the earthquake. The path of mind since the morning. ③ The middle-aged generation who supported the reconstruction.
33. 2000/01/12/Wed/23:00 NHK Ch.2 NHK News 11 'Five years since the earthquake. Composed melody (music) for the poem written by the son who died in the earthquake.'
34. 2000/01/12/Wed/21:30 Sun-TV News EYE f '60% of temporary lease land. Reconstruction Street and land adjustment and improvement (Kukaku Seiri).'
35. 2000/01/13/Thu/05:00 NHK Ch.2 Good Morning Japan 'Prime Minister Obuchi visiting Thailand', 'Now Public Reconstruction Housing,' 'Disaster Prevention training', 'Diagnosis of the seriously injured at the time of disaster'.
36. 2000/01/13/Thu/08:30 NHK CH. 2 Life Hot. 'Five years since the earthquake. Hurt of mind does not heal. The suffering of mother who lost her daughter.'
37. 2000/01/13/Thu/11:30-12:00 NHK Ch.2 Before lunch. 'The Great Hanshin Earthquake and Volunteers'
38. 2000/01/13/Thu/16:00 TV-Osaka. Ch.19 Happy Full score TV. 'Five years since the earthquake. The loneliness at the public reconstruction housings.'

39. 2000/01/13/Thu/17:54 Kansai TV Ch.8 Super News Kansai 'Five years after the earthquake. From a disaster victim's will. Merchandise Loan which was able to be paid back by the life of husband'.
40. 2000/01/13/Thu/18:23 Yomiuri TV Ch.10 News Scramble 'Kizuna (relationship) of the fifth year. Disappearing temporary shelters.'
41. 2000/01/13/Thu/18:30 ABC TV Ch.6 Wide ABCDEsu 'The path of our town reconstruction'.
42. 2000/01/13/Thu/18:30 MBS Mainichi TV Ch.4 Now 'Remember the day 1.17 ④ People shaken at the street and land adjustment and improvement (kukaku seiri).
43. 2000/01/13/Thu/19:00 NHK Ch.2 NHK News 7 'If learning from the Earthquake experience alive? The reality of the building examination of which currently under construction'.
44. 2000/01/13/Thu/19:10-45 NHK Edu. Ch.12 Volunteer 'Special. New starts of the earthquake orphans.'
45. 2000/01/13/Thu/21:00 NHK Ch.2 NHK News 9 'Utilization of Electric Map for the reconstruction after disaster'.
46. 2000/01/13/Thu/23:00 NHK Ch.2 NHK News 11 'New methods for diagnosis of the seriously injured at the time of disaster.'
47. 2000/01/13/Thu/21:30 Sun-TV News EYE f 'The light of hope to KOBE'
48. 2000/01/14/Fri/05:00 NHK Ch.2 Good Morning Japan. 'Five years since the earthquake at the brewery'.
49. 2000/01/14/Fri/08:30 NHK Ch.2 Life Hot. 'Kayo won't be a loser. The five years since the earthquake a girl experienced. Divorce of her parents.'
50. 2000/01/14/Fri/17:54 Kansai TV Ch.8 Super News Kansai 'Lighting with heart. Those who live away from Kobe.'
51. 2000/01/14/Fri/18:23 Yomiuri TV Ch.10 News Scramble 'Memories at the memorial. Requiem for 6432 people.'
52. 2000/01/14/Fri/18:30 ABC TV Ch.6 Wide ABCDEsu 'Earthquake volunteers'
53. 2000/01/14/Fri/18:30 Mainichi TV MBS Ch.4 Now 'Memorize the day 1.17. Disappearing temporary factories. Difficult restart.'
54. 2000/01/14/Fri/18:00 NHK Ch.2 News Park Kansai 'From the questionnaire survey of earthquake'
55. 2000/01/14/Fri/19:00 NHK Ch.2 NHK News 7 'Weathering the consciousness for disaster prevention. Learn again from the lessons (kyoukun) of the disaster-affected.'
56. 2000/01/14/Fri/21:00 NHK Ch.2 NHK News 9 'Issues of using the mobile phones at the time of earthquake.'
57. 2000/01/14/Fri/21:30-22:30 NHK Ch.2 NHK Special 'Five years The Great Hanshin Earthquake ④ To save one more life. Lessons from the 6,400 deaths. Examine the rescue system.'
58. 2000/01/14/Fri/23:10 NHK Ch.2 NHK News 11 'New movement on implementing earthquake insurances.'
59. 2000/01/14/Fri/23:35-45 NHK Ch.2 Reading Tomorrow 'Earthquake and landslide'
60. 2000/01/14/Fri/25:10-27:30 NHK Ch.2 Learn from the Great Earthquake. Re-broadcasting. 'The lights of city went off'. 25:55 'Two years of a housewife.' 26:40 'From under the broken stuff.'
61. 2000/01/14/Fri/22:00 ABC TV Ch.6 News Station 'If the minds of the earthquake orphans have been taken care and healed? 'Earthquake orphans home Rainbow House'.
62. 2000/01/14/Fri/21:30 Sun TV 'News EYE f 'Five years after the earthquake. For tomorrow'

63. 2000/01/14/Fri/23:30 MBS Mainichi TV Ch.4 Tsukushi Tetsuya News 'From Kobe to Taiwan. What connects the hearts of them'.
64. 2000/01/15/Sat/18:00 NHK Ch.2 Good Morning Japan 'The current situation at a small size public reconstruction housing', 'Five years since the earthquake. Talk by Prof. Hayashi, Kyoto Univ.'
65. 2000/01/15/Sat/09:15 NHK Ch.2 Saturday Hot Wide 'Tanigawa Kouji talk about his earthquake experience', 'How to build a house to sustain an earthquake'
66. 2000/01/15/Sat/18:10-45 NHK Ch.2 Special Feature: Post Earthquake Reconstruction 'The issues to be left at five years since the earthquake: Community and town development, support for older people'
67. 2000/01/15/Sat/25:15-26:05 NHK Ch.2 'Start of earthquake orphans' (re-broadcast)
68. 2000/01/15/Sat/26:05-27:25 NHK Ch.2 'Run 1,000 km the Silk Road over the earthquake'
69. 2000/01/15/Sat/25:35-26:45 Yomiuri TV Ch.10 Earthquake Special Feature 'EARTHQUAKE' the decision of a child. Lessons from ten seconds that divided lives.
70. 2000/01/16/Sun/07:45-08:28 NHK Ch.2 Earthquake Special. Mr. Tamura, NHK, Kobe, Mr. Kaihara, Hyogo Prefecture Governor, Mr. Kazuto Uchihashi, Economist, Ms. Hiroko Kuroda, Nurse, Representative of Shimin Kikin KOBE (Citizen Fund: KOBE), Mr. Yoishiro Fujiyoshi, NHK Editorial Writer
71. 2000/01/16/Sun/10:00-11:00 MBS Mainich TV Ch.4 Special earthquake program 'With... Kizuna to the world: devoted love of the earthquake experienced to rescue those in the front line of hunger and poverty'.
72. 2000/01/16/Sun/11:00-12:00 TV Osaka Ch.19 'Five years since the earthquake. Looking for the house of healing. Now Kizuna to the lonely heart which has been closed.'
73. Part of the program is re-broadcast of a part of 37. 2000/01/13/Thu/16:00 TV-Osaka. Ch.19 Happy Full score TV. 'Five years since the earthquake. The loneliness at the public reconstruction housings.'
74. 2000/01/16/Sun/11:30-12:00 NHK Ch.2 New Asia. Turkey Great Earthquake.
75. 2000/01/16/Sun/13:00-14:55 Kansai TV Ch.8 'Examine The Great Hanshin Earthquake. What the five years at the disaster affected area question. The time the former Prime Minsiter, Murayama, talk. Those who cannot return to their home town.
76. 2000/01/16/Sun/15:30-17:30 ABC TV Asahi TV. Ch.6 'Eire (yell) 1.17 Rebirth. The wisdom of Kobe'. Lessons from the temporary shelter housings. Town development. Take a photo of the disaster-affected area. Where the public support goes. The evolution of volunteers. Songs of hearts.
77. 2000/01/16/Sun/17:30-18:00 TV Osaka Ch.19 Super mother of Taiwan disaster-affected.
78. 2000/01/16/Sun/18:00- (30 hours) NHK Ch.2 'Earthquake five years'. Support each other. Thirty hours till midnight of 17 Jan to connect the disaster affected areas. Questionnaire to 3500 people.
79. 2000/01/16/Sun/18:00 Mainichi TV Ch.4 Special News. Five years since the earthquake. Examine the risk at our feet.
80. 2000/01/16/Sun/24:35 Mainichi TV Ch.2 Image '00 'The disaster affected area in the cold winds. Now the homeless people'.
81. 2000/01/16/Sun/19:00-54 Sun TV Ch.36 The broken life line. How to utilize the learning experience from the Earthquake.
82. 2000/01/16/Sun/22:00-30 Sun TV Ch.36 Kobe Report. 'Five years after the earthquake. Messages to the next generations.'
83. 2000/01/16/Sun/25:15-27:15 Sun TV Ch.36 Report of rescue activities at the Turkey and the Taiwan earthquakes. Lessons to the Hanshin areas.

84. 2000/01/16/Sun/27:15-29:30 Sun TV Ch.36 The news of the past year. Re-broadcast the EYE f news of the past year. Temporary shelters. Public Reconstruction housings. Street shops.
85. 2000/01/16/Sun/24:55-25:50 Kansai TV Ch.8 'Visiting the wishes. Messages to leave. The records of those earthquake victims.'
86. 2000/01/16/Sun/25:50- Kansai TV Ch.8 'We will not forget you. Monument with our hearts.'
87. 2000/01/16/Sun/24:25-55 Document '00 Kobe By supporting each other. Ms. Hiroko Kuroda (who was a nurse now is a member of the support network to Hanshin old people and disabled people) has been working as a volunteer. By looking at her activities and the disaster-affected people, we will think what is the real reconstruction. Second of the series.
88. 2000/01/17/Mon/05:30 Sun TV Ch.36 Earthquake special program. 'Looking at the future'. 'Lessons from the earthquake'
89. 2000/01/17/Mon/05:00 NHK Ch.2 Good Morning Japan Earthquake. Continues almost all day earthquake specials in NHK, and several hours at each private television station. Introducing people's five years. Memorial activities starting from early morning, the time of the earthquake. Disaster prevention training. Earthquake in Turkey and Taiwan.
90. 2000/01/17/Mon/12:00-45 NHK Ch.2 The Great Hanshin Earthquake memorial ceremony for those who died. Attendants include Crown Prince Hironimiya and Prime Minister Obuchi. They delivered a speech.
91. 2000/01/17/Mon/21:30-23:00 NHK Ch.2 International Corporation. The earthquakes in Turkey. German and Israeli ways compared with Japanese aids. The earthquake in Taiwan. Kukaku Seiri (street improvement). Introducing the very similar case. Street with shops. The reasons for what each shop/house owner wishes for what. The case is very similar from outside. But the policy process is very different. Compared to Japan, Taiwan takes bottom-up approaches. *Gyousei* (public administration) has not been strong in Taiwan in all other sectors. Private sectors have been the one to take an initiatives. This time, too. Also, they have more Christian churches and those people who have experienced working as volunteers. They have those volunteers of the large number already trained. The context is different, too
92. 2000/01/22/Sat

# Appendix 3-4

Table A3-4.1: *Asahi* newspaper Jan 2000

1. 2000/01/09 Sun. Page 1. 'Rapid ageing at post-earthquake reconstruction housings.' Kobe City run 31.9%. Hyogo Prefecture run 36.9%. Nishinomiya city run 44.2%. Time to evaluate and reconsider the support policy.
2. 2000/01/09/sun. Page 29 'Isolated death and suicides 38 people in the past one year'. You are not alone. Public reconstruction housing. Festival. Hobbies. 'Talk to each other'.
3. 2000/01/10/Mon. Page 1. 'Emergency relief. Society bearable to the earthquake. five years after the Hanshin earthquake. Japan's support to the earthquakes in Colombia, Turkey, Taiwan.
4. 2000/01/10 Mon. pages 28-29. Earthquake Symposium. From Hanshin, Turkey and Taiwan.
5. 2000/01/11 Tue. Page 1. Society bearable to the earthquake. Five years after the Hanshin earthquake. Telecommunication at the time of disaster. The rapid increase of mobile phones and decrease of public phone booths. Will it be able to cope?
6. 2000/01/12/Wed. Page 1. Society bearable to the Earthquake. Five years after the Hanshin earthquake. ③ The disaster weak (Saigai Jakusha). The problems and issues in normal daily life were carved out.
7. 2000/01/13/Thu. Page 1. Society bearable to the earthquake. Five years after the Hanshin earthquake. Enterprise Company Disaster Prevention. Continuation of the business and work is also the social responsibility. Power plant. The case of Taiwan earthquake. What was learning experience from Kobe. Comparison of the two earthquakes.
8. 2000/01/14/Fri. Page 1. Society bearable to the earthquake. Five years after the Hanshin earthquake. Towards the stronger housings and stronger town. Lessons learned from the collapse'. Public money was invested. Rules of the citizens.
9. 2000/01/14/Fri Page 1. 'Disaster-affected children. New cases development are reported. Post-traumatic Stress Disorder (PTSD). Hyogo Prefecture Education Committee survey reports that some came to surface after a long period of time.
10. 2000/01/14/Fri Page 29. Interview with Hyogo Governor Kaihara. 'To realize the support system of housing.'
11. 2000/01/15/Sat Page.1 '40% of the houses of total destroyed receive no subsidies. The system to support the disaster-affected. The barriers such as the limitation by the annual income. Low application in urban areas.
12. 2000/01/15/Sat Page 4. International Conference: Examine the Earthquake. 'Mission to make use of and convey the lessons learnt'.
13. 2000/01/15/Sat Page 33. Local pages introduce some memorial events of the earthquake
14. 2000/01/16/Sun Page 1. 'The Great Hanshin Earthquake. Tomorrow. The exact fifth year. Public reconstruction housings. The deepening sense of isolation. Survey of 500 households shows that more than 10% want to move out. The manner is lowering. The complains are increasing.
15. 2000/01/16/Sun Page 3, 4, 38.
16. 2000/01/16/Sun Page 39 80% of the old age disaster-affected go to a hospital regularly. Aftershock to body. Mind. Money. Old age. Survey at the public reconstruction housings. *Gyousei* support. 'Consider more'.
17. 2000/01/16/Sun Pages 19-24 Earthquake special
18. Page 19. Buddhist priest who is a famous writer.

Page 20-21. Mr. Yusuke Kishi (41) writer. 'Always next to risk' Ms. Yukiko Touhata (58) Kobe city Higashi Nada ward. Office woman, who lost her 21 year old only daughter at the earthquake. 'Crying all the time. No'.

Page 22. Photos. Turkey. Taiwan. Temporary shelters sent from Kobe to Taiwan.

Page 23. To Old People, neighbours are too distant. The result of *Asahi* News Survey of 500 people at the public reconstruction housings.

Page 24. For the day that will come sometime. At home, sleep at the edge of the room.

2000/01/17/Mon Page 1. Today Fifth Year The Great Hanshin Earthquake. Lessons learnt. Share with the world and heritage. International Symposium. *Gyousei* and *chiiki* (local) to *Renkei* (collaborate). Suggest examine conference to establish the system for future international cooperation.

Photo. Candle service of 6432 candles. (the number of death toll) at 05:46 the time of the Earthquake. 'Let's move on forward.'

19. 2000/01/17/Mon Pages, 2,3,21, 38-39. Pages 2-3 Panel Discussion
20. 2000/01/17/Mon Evening. Earthquake memorial ceremony.

Transcripts of Messages of Crown Prince and of Prime Minister Obuchi.

## Appendix 3-5-1: TV coverage of Earthquake by station, time and type of program for January 1999 and 2000

Table A3-5-1.1 follows.

Table A3-5-1.1: TV coverage of earthquake by station, time, type of program for January 1999

| Station | News a.m. | News 6 p.m. | News 9 p.m. | News Late | Special Feature | Documentary | Time (min) on elderly | Ratio |
|---|---|---|---|---|---|---|---|---|
| NHK 2 | | 11. TemHouse | | 12. Orphans | 14. Close up Gendai. Lonely old people (30) | 10 (25) | 25 | 105/330 |
| | | 12. | | PTSD | 17. (50) | 15 (25) | 30 | |
| | | 13. Small enterprises. Out of Pref. | | | 17. (50) | 16.Volunteer (30) | 50 | |
| | | 14. Volunteer | | | 18 Reading Tomorrow | 12 (30) | | |
| | | Isolated death | | | (10) | 14NGO (10) | | |
| | | 15. | | | | | | |
| | | 17. Gyousei | | | | | | |
| NHK Edu 12 | | | | | | 14(30) | | 0/30 |
| NHK Satellite BS | | | | | 14 (20) | | | 90 |
| | | | | | 17 (90) Forum. Prof. Konishi | | 90 | 170 |
| | | | | | 17 (60) | | | |
| Mainichi MBS 4 | | 11. Small enterprises. | | 11 (10) | 17 (60) | 17 (60). | 15 | 115/225 |
| | | 12. Factory | | 14 Kuwahara (10) | | Dementia. Home care | 60 | |
| | | 13. Group home LSA | | 15 (15) | | | 10 | |
| | | Kuwahara | | | | | 10 | |
| | | 14. Kizuna Collective House | | | | | | 10 |
| | | 15. | | | | | 10 | |
| | | 17 | | | | | | |
| | | 18. Charity Walk | | | | | | |

Table A3-5-1: continued

| Station | News | | | | Special Feature | Documentary | Time (min) on elderly | Ratio |
|---|---|---|---|---|---|---|---|---|
| | a.m. | 6 p.m. | 9 p.m. | Late | | | | |
| Asahi ABC 6 | | 11. LSA Kuwahara(10) | 18. *Gyosei* | 13 (15) | | 16 (60) | 10 | 85/ |
| | | 12. | | | | | | 60 |
| | | 13. | | 15. Orphans | | | 15 | |
| | | 14. | | | | | 60 | |
| | | 15. Temp was better. | | | | | | |
| | | 18. Disaster prevention. SanFrancisco | | | | | | |
| Kansai 8 | | 11. Temp House | | | | | | 10/ |
| | | 12. Shops | | | | | | 60 |
| | | Temp House | | | | | | |
| | | 13. Barrier Free | | | | | 10 | |
| | | 14. Economic depression | | | | | | |
| | | 15. Memorial | | | | | | |
| | | LSA | | | | | | |
| | | 18. | | | | | | |
| Yomiuri 10 | | 11. Gyosei | 15. Temp House (10) | 10(10/30) | 17 (95) | 17 Kuroda(30) | 10 | 75/ |
| | | 12. Temp House | | | | 17 (65) | 10 | 325 |
| | | Murder of father | | | | 17 (30) | 5 | |
| | | 13. Orphans. | | | | | 10 | |
| | | Old people (5) | | | | | 30 | |
| | | 14. Housing and old people | | | | | | 10 |
| | | 18. | | | | | | |

| | | | | | |
|---|---|---|---|---|---|
| TV Osaka 19 | | | 15(10) | | 0/10 |
| Sun TV 36 | 10(30)<br>11.<br>12. Temp with care for old(10)<br>13 Conflict with Gyōsei (10).<br>14. Shops | Eye-f(30)<br>8.Temp H.<br>PRH<br>1<br>11 Old people<br>12 Temp relocation<br>13<br>14<br>15<br>16. (re broadcast the selection of the past one year.(240)<br>18. Volunteer | 17(60)<br>17(30) | 30<br>30<br>10 | 70/<br>370 (+240) |
| Total | | | | 550 (33%) | 550/1675 (+?/240) |

Table A3-5-1.2: TV coverage of earthquake by station, time, type of program for January 2000

| Station | News: a.m. | News: 6 p.m. | News: 9 p.m. | News: Late | Special Feature | Documentary | Time (min) on elderly | Ratio |
|---|---|---|---|---|---|---|---|---|
| NHK 2 | 11. | 11. DP | | 15. Emergency Disaster Relief. Medical. Telephone Insurance (130) | 9. moring(15) Geology | 14.8:30-(30) Child. Parents' Divorce | 5 | 60/ |
| | 12. Disaster Prevention(DP) | 12. DP | 11. Highway | | 11. Gyosei(10) | 16. Turkey | 35 | 2335 |
| | 13. DP | 13. | 12. DP | | 12. CloseUP Gendai mall enterprises(30) | | 20 | |
| | 14. Beer | 14.NHK Survey result | | 13. DP | | 13. Before Lunch. Volunteer (30) | | |
| | 15. PRH | 15. | 14. Survey Result | | | 14. Emergency Relief (10) | | |
| | 17. | 17. | | | 15. Old (35)<br>16. Old (20/45)<br>16:18:00-17(30Hrs) | | | |
| NHK Edu 12 | | | | | Series of 3:<br>10.Children(45)<br>11.Old (45)<br>12. MiddleAge (45)<br>12(30)<br>13 Volunteer(30) | | 45 | 45/<br>195 |
| NHK Satellite BS | | | | Kokoro (heart) (60) Mother who lost her son | | | 0 | 0/60 |

| | | | | | | |
|---|---|---|---|---|---|---|
| Mainichi MBS 4 | 7. Out Pref.<br>10.<br>11.<br>12.<br>13. KukakuSeiri<br>14. Remember<br>16. | 14. Taiwan | | 16. Kizuna to World (60) | | 0/140 |
| Asahi ABC 6 | 7.<br>10.<br>11. Life saving dog<br>12. Housing(5/10)<br>13.<br>14.V olunteer | 14. Orphans (10) | | 16 (120) | 5 | 5/190 |
| Kansai 8 | 7. Tunagari. life reconstruction<br>10. Shops<br>11. Those who lost children<br>12. Kobe City Mayor. Old people. Housing<br>13<br>14. Kokoro (Heart) | 16 (120) | 16 (115) | | 10 | 10/295 |
| Yomiuri 10 | 7.<br>10. Shops<br>11. Those who lost children<br>12. Children<br>13. Kizuna<br>14. Memorial | | 15 (70) | 9 (30)<br>16 (30) | 30 | 30/195 |

Table A3-5-1.2: continued

| Station | News | | | | Special Feature | Documentary | Time (min) on elderly | Ratio |
|---|---|---|---|---|---|---|---|---|
| | a.m. | 6 p.m. | 9 p.m. | Late | | | | |
| TV Osaka 19 | | | | | 13. 16:00 100 full score. (10)<br>16. Kizuna (60)<br>16. 100 Full score (10)<br>16. Taiwan (30) | | | 20/110 |
| Sun TV 36<br>30/180(+?/375) | | 16. Telephone<br>17. Memorial events. | Eye-f(30)16<br>7. Life Reconstruction<br>10. Disaster Prevention<br>11.DP<br>12. KukakuSeiri<br>13. Light of hope | 16 Re broadcast the selection of the past one year (375 min) | | | 30 | |
| Total | | | | | | | 200 (8.4%)<br>(37%) | 200/2335<br>(+?/375)<br>if w/oNHK<br>30hrs,<br>200/535 |

# Appendix 3-5-2: QSR NVivo project

Diagram A3-5-2.1: Sets created in NVivo

---

- Media
- Haruyama TSH
- Natsu Aki PRH
- Fuyuyama PRH

---

Diagram A3-5-2.2: Tree nodes created in NVivo

---

- TV station
  - NHK
  - Mainichi
  - Asahi
  - Kansai
  - Yomiuri
  - NHK Education
  - NHK BS Satellite
  - Sun-TV
  - TV-Osaka
- Age
  - Old people
  - Young people
- Gender
  - Men
  - Women
- Housing Type
  - *Nagaya*
  - Temporary Shelter Housing
  - Public Reconstruction Housing
  - General Housing
  - Group Home
  - Collective House
  - Silver Housing
- Actor
  - Family
  - Government
  - *Gyōsei*
  - Housewife

*Jichikaicho*
LSA
Media
*Minsei Iin*
Neighbours
NGOs
Physician
Prefecture Government
Staff
University
Volunteer
*Jichikai* (self-governing body)
Public health nurse
Department of Health
Director of Department of Health
*Jumin* (residents)

Health Issues
Alcohol dependency
Bedridden
Demented
Depression
Health
Health promotion
PTSD

---

# Appendix 3-5-3: List of survey findings related to being old from each year.

## FY 1996

1. Among those in temporary shelters, 42.9% of people aged 80 and over said that their health was 'not good' or 'not so good'.
2. Among those in general housing, 37.2% of people aged in their 70s and 29.6% in their 80s said that their health was 'not good' or 'not so good'.
3. During one month, 'easy to get tired', 'shoulder pain' and 'headache', were found to be the most common health problems for residents in their 50s, out of a total of eight symptoms. People in their 60s reported 'hip back pain' and 'joint pain'.
4. Residents aged in their 70s reported the highest proportion of illness (84.1%), followed by those in their 80s (72.25%). Younger people reported less illness.
5. GHQ30 mental problems were most common among residents in their 50s, while PTSS-10 problems were most common among those in their 60s.

## FY 1997

1. Residents aged in their 50s (54%) were the most common of temporary shelter residents to say that their health was 'not good' or 'not so good'.
2. 'Joint pain' was reported most by residents aged 60 and older.
3. The older the resident, the higher proportion reported illness (13.5% of people aged in their 10s, 83.3% of people in their 80s and 81% of people in their 60s.
4. Alcohol problems were most common among people in the 50s (14.5%), followed by those in their 40s.
5. Residents aged in their 50s (42%) were the most common in public reconstruction housing to say that their health was 'not good' or 'not so good'. The older the resident, the more likely they were to report health problems.
6. 'Hip and back pain' was most commonly reported among residents aged in their 40s and older.
7. The older the resident, the more likely they were to report illness (15.5% of those aged in their 10s, 79.6% in the 80s and older). Among those in their 60s and older, 77.4% reported illness.
8. Residents aged in their 40s and 50s were found to have the most mental health problems on both the PTSS-10 and GHQ30 measures.
9. Alcohol problems were most commonly reported among people in their 40s (11.0%), followed by those in their 50s.
10. The order the resident, the higher their expectations for public health and welfare services.
11. Among general housing residents, those aged in their 70s (35%) were the most likely to report that their health was 'not good' or 'not so good', followed by those in their 80s and older (30.9%), their 50s (25.5%), their 40s (24.5%), their 60s (23.7%), their 10s (16.1%), their 30s (11.2%) and their 20s (10.6%).
12. 'Hip and back pain' was most commonly reported by residents in their 40s and over.
13. Residents in their 70s (80.4%) were most likely to report illness, followed by those aged 80 and over (72.1%).
14. Alcohol problems were most commonly reported by residents aged in their 40s (8.8%), followed by those in their 50s.

**FY 1998**

1. The IES-R measure of mental health problems reported the highest score among residents aged in their 60s (45.4%).
2. All age groups reported lower than target consumption of 'vegetables'. Residents aged in their 80s also reported lower than target consumption of 'beans, fish, meat, egg'.
3. In public reconstruction housing, the older the resident, the more likely they were to live alone. 39.4% of those aged 65 and older were living alone.
4. Residents aged in their 80s (42.4%) were most likely to report that their health was 'not good' or 'not so good', followed by those aged in their 50s (38.4%).
5. Residents aged in their 70s (60.7%) were the most likely to seek a medical check.
6. Mental health problems were more common among residents in aged their 50s (34.2%) and early 60s (34.4%), as measured by the IES-R test.
7. Residents in their 40s were most likely to be reported as having alcohol related problems.

# Appendix 3-6: Sample of the Hyogo Prefecture post-earthquake Health Survey

Table A3-6.1: Age and sex composition of TSH residents

| | 1996 | | 1997 | | 1998 | |
|---|---|---|---|---|---|---|
| | Number | % | Number | % | Number | % |
| *Sex* | | | | | | |
| Men | 2302 | 43.3 | 1547 | 42.5 | 338 | 47.5 |
| Women | 3013 | 56.7 | 2097 | 57.5 | 429 | 52.5 |
| *Age* | | | | | | |
| 0–19 | 177 | 3.3 | 89 | 2.5 | 19 | 2.3 |
| 20–29 | 476 | 9.0 | 260 | 7.1 | 56 | 6.9 |
| 30–39 | 403 | 7.6 | 221 | 6.1 | 62 | 7.6 |
| 40–49 | 671 | 12.6 | 387 | 10.7 | 100 | 12.2 |
| 50–59 | 933 | 17.6 | 635 | 17.4 | 186 | 22.8 |
| 60–69 | 1397 | 26.3 | 1027 | 28.2 | 97 | 11.9 |
| 70–79 | 862 | 16.2 | 708 | 19.4 | 111 | 13.6 |
| 80–89 | 336 | 6.3 | 285 | 7.8 | 134 | 16.4 |
| 90+ | 25 | 0.5 | 32 | 0.9 | 49 | 6.0 |
| No answer | 35 | 0.7 | 0 | 0.0 | 3 | 0.4 |
| Total | 5315 | 100.0 | 3644 | 100.0 | 817 | 100.0 |

Table A3-6.2: Age and sex composition of PRH residents

| | 1997 | | 1998 | |
|---|---|---|---|---|
| | Number | % | Number | % |
| *Sex* | | | | |
| Men | 1319 | 41.7 | 2577 | 41.2 |
| Women | 1846 | 58.3 | 3671 | 58.5 |
| *Age* | | | | |
| 0–19 | 84 | 2.6 | 130 | 2.1 |
| 20–29 | 348 | 11.0 | 597 | 9.6 |
| 30–39 | 264 | 8.3 | 493 | 7.9 |
| 40–49 | 264 | 8.3 | 521 | 8.3 |
| 50–59 | 464 | 29.1 | 962 | 15.4 |
| 60–69 | 911 | 14.4 | 1784 | 28.6 |
| 70–79 | 600 | 19.0 | 1241 | 19.9 |
| 80–89 | 206 | 6.5 | 495 | 7.9 |

Table A3-6.2: continued

| | 1997 | | 1998 | |
|---|---|---|---|---|
| | Number | % | Number | % |
| 90+ | 24 | 0.8 | | |
| No answer | 0 | 0.0 | 25 | 0.4 |
| Total | 3165 | 100.0 | 6248 | 100.0 |

*80–89 and 90+ were combined in 1998.

Table A3-6.3: Age and sex composition of GH residents

| | 1996 | | 1997 | |
|---|---|---|---|---|
| | Number | % | Number | % |
| *Sex* | | | | |
| Men | 397 | 44.0 | 441 | 42.9 |
| Women | 505 | 56.0 | 588 | 57.1 |
| *Age* | | | | |
| 0–19 | 22 | 2.4 | 31 | 3.0 |
| 20–29 | 104 | 11.5 | 114 | 11.1 |
| 30–39 | 140 | 15.5 | 161 | 15.6 |
| 40–49 | 129 | 14.3 | 147 | 14.3 |
| 50–59 | 148 | 16.4 | 137 | 13.3 |
| 60–69 | 182 | 20.2 | 228 | 22.2 |
| 70–79 | 113 | 12.5 | 143 | 13.9 |
| 80–89 | 44 | 4.9 | 59 | 5.7 |
| 90+ | 10 | 1.1 | 9 | 0.9 |
| No answer | 10 | 1.1 | 0 | 0.0 |
| Total | 902 | 100.0 | 1029 | 100.0 |

*60–64 and 65–69 were combined in 1997.

Table A3-6.4: Age composition of Hyogo Health Survey for TSH residents by sex in 1996

| | Men | | Women | | Total | |
|---|---|---|---|---|---|---|
| TSH 1996 | Number | % | Number | % | Number | % |
| 0–19 | 85 | 48.0 | 92 | 52.0 | 177 | 100.0 |
| 20–29 | 225 | 47.3 | 251 | 52.7 | 476 | 100.0 |
| 30–39 | 201 | 49.9 | 202 | 50.1 | 403 | 100.0 |
| 40–49 | 335 | 49.9 | 336 | 50.1 | 671 | 100.0 |
| 50–59 | 390 | 41.8 | 543 | 58.2 | 933 | 100.0 |
| 60–64 | 334 | 45.6 | 398 | 54.4 | 732 | 100.0 |
| 65–69 | 282 | 42.4 | 383 | 57.6 | 665 | 100.0 |
| 70–79 | 321 | 37.2 | 541 | 62.8 | 862 | 100.0 |
| 80+ | 115 | 31.9 | 246 | 68.1 | 361 | 100.0 |
| No answer | 14 | 40.0 | 21 | 60.0 | 35 | 100.0 |
| Total | 2302 | 43.3 | 3013 | 56.7 | 5315 | 100.0 |

* Age group 60–69 is divided into 60–64 and 65–69.

Table A3-6.5: Sex composition of TSH residents by age group in 1996

| | Men | | Women | | Total | |
|---|---|---|---|---|---|---|
| TSH 1996 | Number | % | Number | % | Number | % |
| 0–19 | 85 | 3.7 | 92 | 3.1 | 177 | 3.3 |
| 20–29 | 225 | 9.8 | 251 | 8.3 | 476 | 9.0 |
| 30–39 | 201 | 8.7 | 202 | 6.7 | 403 | 7.6 |
| 40–49 | 335 | 14.6 | 336 | 11.2 | 671 | 12.6 |
| 50–59 | 390 | 16.9 | 543 | 18.0 | 933 | 17.6 |
| 60–64 | 334 | 14.5 | 398 | 13.2 | 732 | 13.8 |
| 65–69 | 282 | 12.3 | 383 | 12.7 | 665 | 12.5 |
| 70–79 | 321 | 13.9 | 541 | 18.0 | 862 | 16.2 |
| 80+ | 115 | 5.0 | 246 | 8.2 | 361 | 6.8 |
| No answer | 14 | 0.6 | 21 | 0.7 | 35 | 0.7 |
| Total | 2302 | 100.0 | 3013 | 100.0 | 5315 | 100.0 |

* Age group 60–69 is divided into 60–64 and 65–69.

Table A3-6.6: Age composition of Hyogo Health Survey for TSH residents by sex in 1997

| | Men | | Women | | Total | |
|---|---|---|---|---|---|---|
| TSH 1997 | Number | % | Number | % | Number | % |
| 0–19 | 45 | 50.6 | 44 | 49.4 | 89 | 100.0 |
| 20–29 | 124 | 47.7 | 136 | 52.3 | 260 | 100.0 |
| 30–39 | 117 | 52.9 | 104 | 47.1 | 221 | 100.0 |
| 40–49 | 191 | 49.4 | 196 | 50.6 | 387 | 100.0 |
| 50–59 | 289 | 45.5 | 346 | 54.5 | 635 | 100.0 |
| 60–64 | 210 | 41.5 | 296 | 58.5 | 506 | 100.0 |
| 65–69 | 227 | 43.6 | 294 | 56.4 | 521 | 100.0 |
| 70–79 | 242 | 34.2 | 466 | 65.8 | 708 | 100.0 |
| 80+ | 102 | 32.2 | 215 | 67.8 | 317 | 100.0 |
| Total | 1547 | 42.5 | 2097 | 57.5 | 3644 | 100.0 |

* Age group 60–69 is divided into 60–64 and 65–69.

Table A3-6.7: Sex composition of Hyogo Health Survey for TSH residents by age group in 1997

| | Men | | Women | | Total | |
|---|---|---|---|---|---|---|
| TSH 1997 | Number | % | Number | % | Number | % |
| 0–19 | 45 | 2.9 | 44 | 2.1 | 89 | 2.5 |
| 20–29 | 124 | 8.0 | 136 | 6.5 | 260 | 7.1 |
| 30–39 | 117 | 7.6 | 104 | 5.0 | 221 | 6.1 |
| 40–49 | 191 | 12.3 | 196 | 9.3 | 387 | 10.6 |
| 50–59 | 289 | 18.7 | 346 | 16.5 | 635 | 17.4 |
| 60–64 | 210 | 13.6 | 296 | 14.1 | 506 | 13.9 |
| 65–69 | 227 | 14.7 | 294 | 14.0 | 521 | 14.3 |
| 70–79 | 242 | 15.6 | 466 | 22.2 | 708 | 19.4 |
| 80+ | 102 | 6.6 | 215 | 10.3 | 317 | 8.7 |
| Total | 1547 | 100.0 | 2097 | 100.0 | 3644 | 100.0 |

* Age group 60–69 is divided into 60–64 and 65–69.

Table A3-6.8: Age composition of Hyogo Health Survey for TSH residents by sex in 1998

| | Men | | Women | | Total | |
|---|---|---|---|---|---|---|
| **TSH 1998** | **Number** | **%** | **Number** | **%** | **Number** | **%** |
| 0–19 | 9 | 47.6 | 10 | 52.6 | 19 | 100.0 |
| 20–29 | 26 | 46.4 | 30 | 53.6 | 56 | 100.0 |
| 30–39 | 40 | 64.5 | 22 | 35.5 | 62 | 100.0 |
| 40–49 | 52 | 52.0 | 48 | 48.0 | 100 | 100.0 |
| 50–59 | 107 | 57.5 | 79 | 42.5 | 186 | 100.0 |
| 60–64 | 46 | 47.4 | 51 | 52.6 | 97 | 100.0 |
| 65–69 | 50 | 45.0 | 61 | 55.0 | 111 | 100.0 |
| 70–79 | 41 | 30.6 | 93 | 69.4 | 134 | 100.0 |
| 80+ | 15 | 30.6 | 34 | 69.4 | 49 | 100.0 |
| No answer | 2 | 66.7 | 1 | 33.3 | 3 | 100.0 |
| Total | 388 | 47.5 | 429 | 52.5 | 817 | 100.0 |

* Age group 60–69 is divided into 60–64 and 65–69.

Table A3-6.9: Sex composition of Hyogo Health Survey for TSH residents by age group in 1998

| | Men | | Women | | Total | |
|---|---|---|---|---|---|---|
| **TSH 1998** | **Number** | **%** | **Number** | **%** | **Number** | **%** |
| 0–19 | 9 | 2.3 | 10 | 2.3 | 19 | 2.3 |
| 20–29 | 26 | 6.7 | 30 | 7.0 | 56 | 6.9 |
| 30–39 | 40 | 10.3 | 22 | 5.1 | 62 | 7.6 |
| 40–49 | 52 | 13.4 | 48 | 11.2 | 100 | 12.2 |
| 50–59 | 107 | 27.6 | 79 | 18.4 | 186 | 22.8 |
| 60–64 | 46 | 11.9 | 51 | 11.9 | 97 | 11.9 |
| 65–69 | 50 | 12.9 | 61 | 14.2 | 111 | 13.6 |
| 70–79 | 41 | 10.6 | 93 | 21.7 | 134 | 16.4 |
| 80+ | 15 | 3.9 | 34 | 7.9 | 49 | 6.0 |
| No answer | 2 | 0.5 | 1 | 0.2 | 3 | 0.4 |
| Total | 388 | 100.0 | 429 | 100.0 | 817 | 100.0 |

* Age group 60–69 is divided into 60–64 and 65–69.

Table A3-6.10: Age composition of Hyogo Health Survey for PRH residents by sex in 1997

| | Men | | Women | | Total | |
|---|---|---|---|---|---|---|
| **PRH 1997** | Number | % | Number | % | Number | % |
| 0–19 | 42 | 50.0 | 42 | 50.0 | 84 | 100.0 |
| 20–29 | 156 | 44.8 | 192 | 55.2 | 348 | 100.0 |
| 30–39 | 126 | 47.7 | 138 | 52.3 | 264 | 100.0 |
| 40–49 | 126 | 47.7 | 138 | 52.3 | 264 | 100.0 |
| 50–59 | 168 | 36.2 | 296 | 63.8 | 464 | 100.0 |
| 60–64 | 194 | 42.6 | 261 | 57.4 | 455 | 100.0 |
| 65–69 | 220 | 48.2 | 236 | 51.8 | 456 | 100.0 |
| 70–79 | 210 | 35.0 | 390 | 65.0 | 600 | 100.0 |
| 80+ | 77 | 33.5 | 153 | 66.5 | 230 | 100.0 |
| Total | 1319 | 41.7 | 1846 | 58.3 | 3165 | 100.0 |

* Age group 60–69 is divided into 60–64 and 65–69.

Table A3-6.11: Sex composition of Hyogo Health Survey for PRH residents by age in 1997

| | Men | | Women | | Total | |
|---|---|---|---|---|---|---|
| **PRH 1997** | Number | % | Number | % | Number | % |
| 0–19 | 42 | 3.2 | 42 | 2.3 | 84 | 2.6 |
| 20–29 | 156 | 11.8 | 192 | 10.4 | 348 | 11.0 |
| 30–39 | 126 | 9.6 | 138 | 7.5 | 264 | 8.3 |
| 40–49 | 126 | 9.6 | 138 | 7.5 | 264 | 8.3 |
| 50–59 | 168 | 12.7 | 296 | 16.0 | 464 | 14.7 |
| 60–64 | 194 | 14.7 | 261 | 14.1 | 455 | 14.4 |
| 65–69 | 220 | 16.7 | 236 | 12.8 | 456 | 14.4 |
| 70–79 | 210 | 15.9 | 390 | 21.1 | 600 | 19.0 |
| 80+ | 77 | 5.8 | 153 | 8.3 | 230 | 7.3 |
| Total | 1319 | 100.0 | 1846 | 100.0 | 3165 | 100.0 |

* Age group 60–69 is divided into 60–64 and 65–69.

Table A3-6.12: Age composition of Hyogo Health Survey for PRH residents by sex in 1998

| | Men | | Women | | Total | |
|---|---|---|---|---|---|---|
| **PRH 1998** | Number | % | Number | % | Number | % |
| 0–19 | 52 | 40.0 | 78 | 60.0 | 130 | 100.0 |
| 20–29 | 261 | 43.7 | 336 | 56.3 | 597 | 100.0 |
| 30–39 | 238 | 48.3 | 255 | 51.7 | 493 | 100.0 |
| 40–49 | 244 | 46.8 | 277 | 53.2 | 521 | 100.0 |
| 50–59 | 381 | 39.6 | 581 | 60.4 | 962 | 100.0 |
| 60–64 | 364 | 43.5 | 473 | 56.5 | 837 | 100.0 |
| 65–69 | 416 | 43.9 | 531 | 56.1 | 947 | 100.0 |
| 70–79 | 462 | 37.2 | 779 | 62.8 | 1241 | 100.0 |
| 80+ | 151 | 30.5 | 344 | 69.5 | 495 | 100.0 |
| No answer | 8 | 32.0 | 17 | 68.0 | 25 | 100.0 |
| Total | 2577 | 41.2 | 3671 | 58.8 | 6248 | 100.0 |

* Age group 60–69 is divided into 60–64 and 65–69.

Table A3-6.13: Sex composition of Hyogo Health Survey for PRH residents by age in 1998

| | Men | | Women | | Total | |
|---|---|---|---|---|---|---|
| **PRH 1998** | Number | % | Number | % | Number | % |
| 0–19 | 52 | 2.0 | 78 | 2.1 | 130 | 2.1 |
| 20–29 | 261 | 10.1 | 336 | 9.2 | 597 | 9.6 |
| 30–39 | 238 | 9.2 | 255 | 6.9 | 493 | 7.9 |
| 40–49 | 244 | 9.5 | 277 | 7.5 | 521 | 8.3 |
| 50–59 | 381 | 14.8 | 581 | 15.8 | 962 | 15.4 |
| 60–64 | 364 | 14.1 | 473 | 12.9 | 837 | 13.4 |
| 65–69 | 416 | 16.1 | 531 | 14.5 | 947 | 15.2 |
| 70–79 | 462 | 17.9 | 779 | 21.2 | 1241 | 19.9 |
| 80+ | 151 | 5.9 | 344 | 9.4 | 495 | 7.9 |
| No answer | 8 | 0.3 | 17 | 0.5 | 25 | 0.4 |
| Total | 2577 | 100.0 | 3671 | 100.0 | 6248 | 100.0 |

* Age group 60–69 is divided into 60–64 and 65–69.

Table A3-6.14: Age composition of Hyogo Health Survey for GH residents by sex in 1996

| | Men | | Women | | Total | |
|---|---|---|---|---|---|---|
| GH 1996 | Number | % | Number | % | Number | % |
| 0–19 | 9 | 40.9 | 13 | 59.1 | 22 | 100.0 |
| 20–29 | 43 | 41.3 | 61 | 58.7 | 104 | 100.0 |
| 30–39 | 63 | 45.0 | 77 | 55.0 | 140 | 100.0 |
| 40–49 | 59 | 45.7 | 70 | 54.3 | 129 | 100.0 |
| 50–59 | 56 | 37.8 | 92 | 62.2 | 148 | 100.0 |
| 60–64 | 55 | 57.3 | 41 | 42.7 | 96 | 100.0 |
| 65–69 | 38 | 44.2 | 48 | 55.8 | 86 | 100.0 |
| 70–79 | 44 | 38.9 | 69 | 61.1 | 113 | 100.0 |
| 80+ | 26 | 48.1 | 28 | 51.9 | 54 | 100.0 |
| No answer | 4 | 40.0 | 6 | 60.0 | 10 | 100.0 |
| Total | 397 | 44.0 | 505 | 56.0 | 902 | 100.0 |

* Age group 60–69 is divided into 60–64 and 65–69.

Table A3-6.15: Sex composition of Hyogo Health Survey for GH residents by age in 1996

| | Men | | Women | | Total | |
|---|---|---|---|---|---|---|
| GH 1996 | Number | % | Number | % | Number | % |
| 0–19 | 9 | 2.3 | 13 | 2.6 | 22 | 2.4 |
| 20–29 | 43 | 10.8 | 61 | 12.1 | 104 | 11.5 |
| 30–39 | 63 | 15.9 | 77 | 15.2 | 140 | 15.5 |
| 40–49 | 59 | 14.9 | 70 | 13.9 | 129 | 14.3 |
| 50–59 | 56 | 14.1 | 92 | 18.2 | 148 | 16.4 |
| 60–64 | 55 | 13.9 | 41 | 8.1 | 96 | 10.6 |
| 65–69 | 38 | 9.6 | 48 | 9.5 | 86 | 9.5 |
| 70–79 | 44 | 11.1 | 69 | 13.7 | 113 | 12.5 |
| 80+ | 26 | 6.5 | 28 | 5.5 | 54 | 6.0 |
| No answer | 4 | 1.0 | 6 | 1.2 | 10 | 1.1 |
| Total | 397 | 100.0 | 505 | 100.0 | 902 | 100.0 |

* Age group 60–69 is divided into 60–64 and 65–69.

Table A3-6.16: Age composition of Hyogo Health Survey for GH residents by sex in 1997

| | Men | | Women | | Total | |
|---|---|---|---|---|---|---|
| **GH 1997** | **Number** | **%** | **Number** | **%** | **Number** | **%** |
| 0–19 | 15 | 48.8 | 16 | 51.6 | 31 | 100.0 |
| 20–29 | 48 | 42.1 | 66 | 57.9 | 114 | 100.0 |
| 30–39 | 72 | 44.7 | 89 | 55.3 | 161 | 100.0 |
| 40–49 | 74 | 50.3 | 73 | 49.7 | 147 | 100.0 |
| 50–59 | 47 | 34.3 | 90 | 65.7 | 137 | 100.0 |
| 60–64 | 51 | 42.5 | 69 | 57.5 | 120 | 100.0 |
| 65–69 | 47 | 43.5 | 61 | 56.5 | 108 | 100.0 |
| 70–79 | 62 | 43.4 | 81 | 56.6 | 143 | 100.0 |
| 80+ | 25 | 36.8 | 43 | 63.2 | 68 | 100.0 |
| Total | 441 | 42.9 | 588 | 57.1 | 1029 | 100.0 |

* Age group 60–69 is divided into 60–64 and 65–69.

Table A3-6.17: Sex composition of Hyogo Health Survey for GH residents by age in 1997

| | Men | | Women | | Total | |
|---|---|---|---|---|---|---|
| **GH 1997** | **Number** | **%** | **Number** | **%** | **Number** | **%** |
| 0–19 | 15 | 3.4 | 16 | 2.7 | 31 | 3.0 |
| 20–29 | 48 | 10.9 | 66 | 11.2 | 114 | 11.1 |
| 30–39 | 72 | 16.3 | 89 | 15.2 | 161 | 15.6 |
| 40–49 | 74 | 16.8 | 73 | 12.4 | 147 | 14.3 |
| 50–59 | 47 | 10.6 | 90 | 15.3 | 137 | 13.3 |
| 60–64 | 51 | 11.6 | 69 | 11.7 | 120 | 11.7 |
| 65–69 | 47 | 10.6 | 61 | 10.4 | 108 | 10.5 |
| 70–79 | 62 | 14.1 | 81 | 13.8 | 143 | 13.9 |
| 80+ | 25 | 5.7 | 43 | 7.3 | 68 | 6.6 |
| Total | 441 | 100.0 | 588 | 100.0 | 1029 | 100.0 |

* Age group 60–69 is divided into 60–64 and 65–69.

# Appendix A4-1

Table A4-1.1: Surveyed items in the Hyogo Health Survey by each year, 1996–1998

| | 1996 | 1997 | 1998 |
|---|---|---|---|
| Demographic and social indicators: sex, age, address before the earthquake, if has a job or not, the number of family members, family composition | V | V | V |
| Earthquake impact, if any deaths in kinship, home damage | V | V | V |
| Health status, health awareness, health status before the earthquake and that of the past one month, if any illness and its treatment situations | V | V | V |
| Mental Health Status (PTSS-10, GHQ30) | V | V | |
| Mental Health Status (IES-R, Degree of depression) | | | V |
| Any issues in the post-earthquake life (open-ended) | V | | |
| The way to cope with the stress | V | V | |
| What has supported in their post-earthquake life | V | | |
| Home visit service support by public administration and volunteers | | | V |
| Social support | | | V |
| Alcohol-related issues | | V | V |
| Nutritional status, diet, eating habit | | V | V |
| Health and welfare needs | | V | |
| Health care seeking behaviour and situation | | | V |
| Habit (exercise, smoking) | | | V |

# Appendix A4-2: Media analysis

Table A4-2.1: The number of minutes of TV coverage of positive or negative images about different types of housing schemes by TV station, program type (news, special feature, documentary) and year in January of 1999 and 2000.

| | January 1999 | | | | January 2000 | | | |
|---|---|---|---|---|---|---|---|---|
| | TSH(+) | TSH(-) | PRH(+) | PRH(-) | TSH (+) | TSH(-) | PRH(+) | PRH(-) |
| NHK Ch.2 | | | | SF (30) | | | | SF (35) |
| NHKEdu Ch.12 | | | | | | SF (45) | | |
| | | | | | | SF (45)<br>2 cases | | |
| NHK BS Satellite | | | SF (90) | | | | | |
| MBS Mainichi Ch.4 | SF (10) | | N (10)<br>2 cases | N (10)<br>SF (30) | | | | |
| ABC Asahi Ch.6 | | | | N (20)<br>D (10) | | | | |
| Kansai Ch.8 | | | | | | | | N (10) |
| Yomiuri Ch.10 | N (10) Collective House | | N (10) | | N (10)<br>N (10) | | | |
| | | D (30) | | D (30) | D (30) | | | |
| | | | | | | D (30) | | |
| TV Osaka Ch.19 | | | | | | | | D (20) |
| Sun-TV Ch. 36 | | | N (30) | N (30) | | | N (30) | |

N: News, SF: Special Feature, D: Documentary

(+) Positive, (-) Negative

Table A4-2.2: TV coverage of TSH, PRH and GH by station, year, length (minutes) and image.

| TV Station | 1999 | 2000 |
|---|---|---|
| NHK 2 | SF PRH (+ -) 30 (98.12.05)<br>SF PRH (-)30 | SF 35 PRH(-) |
| NHK Edu 12 | | SF PRH(+) TSH(-)<br>45 SF TH (-)<br>2 cases 45 |
| NHK BS | SF PRH (+-) 90 | |
| MBS Mainichi4 | N PRH(+)<br>2 cases 10 | SF TSH(+)<br>10/30 N<br>PRH(-)10<br>SF PRH(-) 30 |
| ABC Asahi 6 | N PRH(-)<br>10 N PRH(-)<br>10D PRH(-) 30 | |
| Kansai 8 | | N PRH(-) 10 |
| Yomiuri 10 | N TSHCollective(+)<br>PRH(-) 10 N<br>PRH(-) 10<br>N TSH(-) 10<br>D PRH(-) TH(-) 30 | D TSH(+ -) 30<br>D PRH(+ -) TSH(-) 30 |
| TV Osaka 19 | | D PRH(-) 10*2 |
| Sun-TV 36 | N PRH(+) 30<br>N PRH(-) 30 | N PRH(+) 30 |

N: News SF: Special Feature D: Documentary

# Maps

Map 1: Map of Japan and surrounding areas

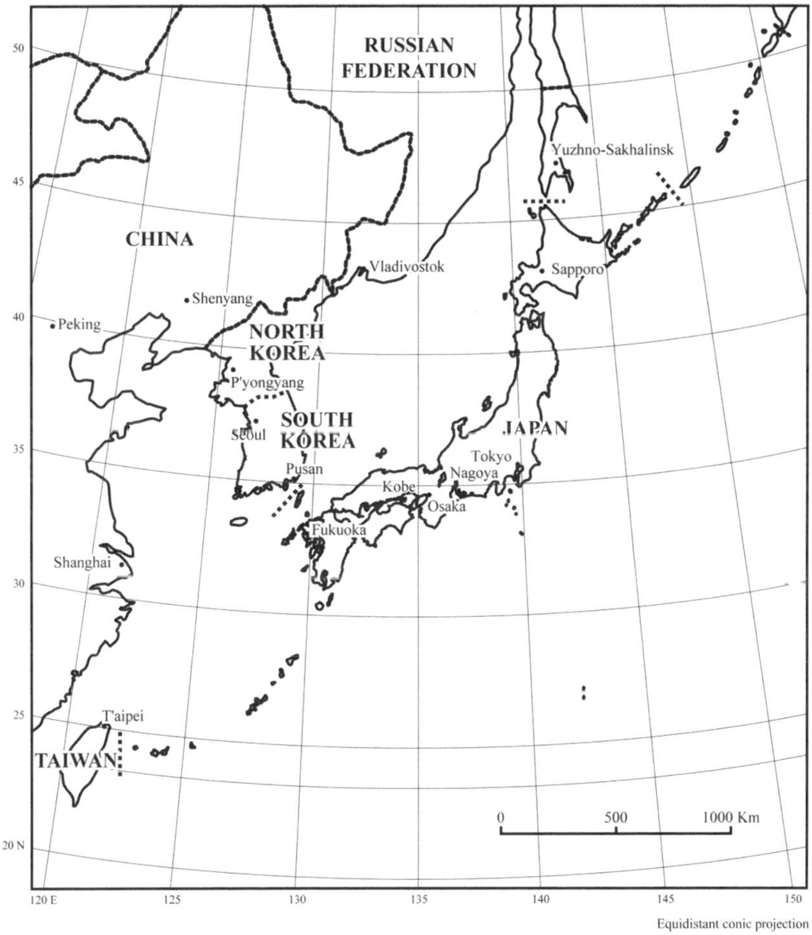


Copyright © MAPIO/DesignExchange Co., Ltd.

Map 2: The number of TSH households

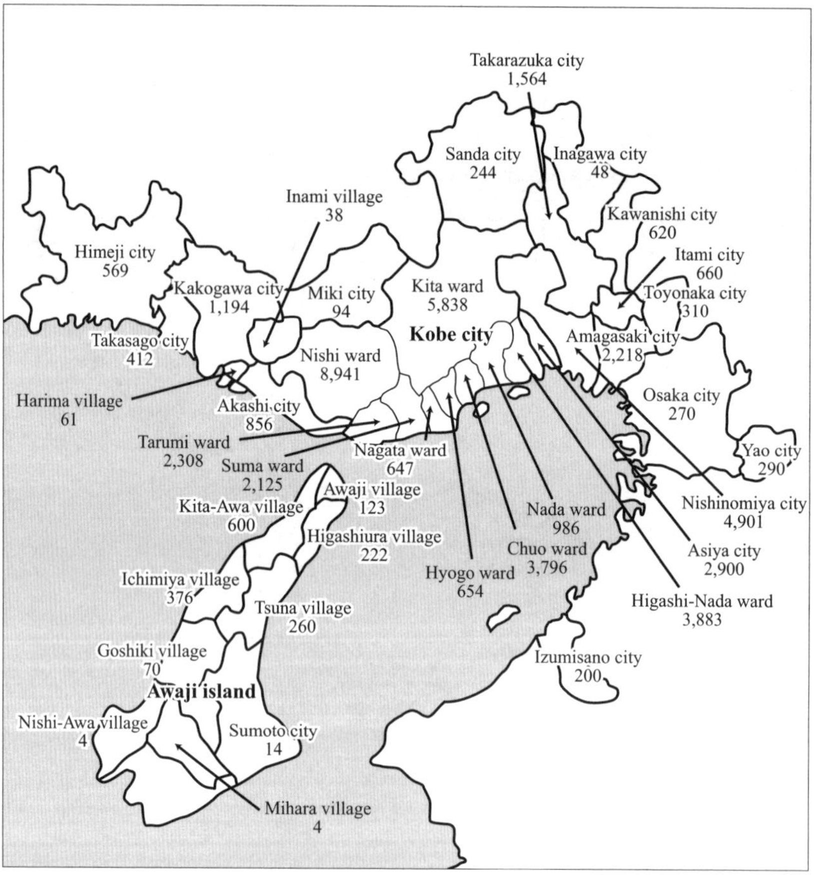


Note: The number of TSH households as of 1 August 1998 (Heinsei 10). The number of households built: 48,300. The current household number: 39,914. The contracted household number: 12, 269. The vacant household number: 27,645.

Map 3: Earthquake zone (more than magnitude 7)

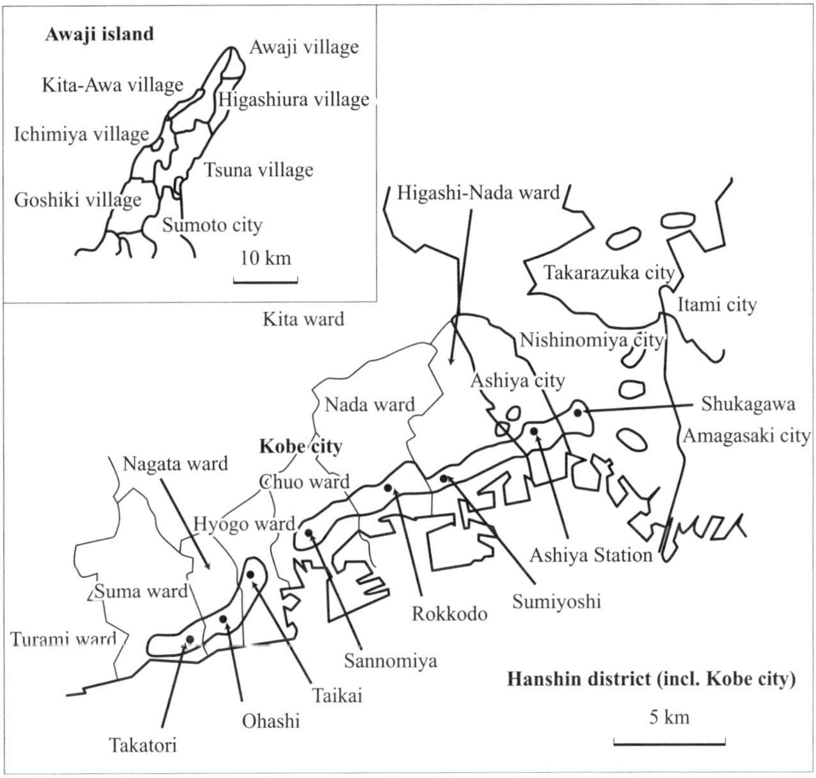


Source: *Shinsai Fukko no Seisaku Kagaku (Policy studies of Post- Earthquake Reconstruction)* Ritsumeikan Daigaku Shinsai Fukko Kenkyu Project (Ritsumeikan University Earthquake Reconstruction Research ProjectÅj (ed), Yuhikaku, 1998, p.92 [Japanese].

Map 4: Map of liquefaction

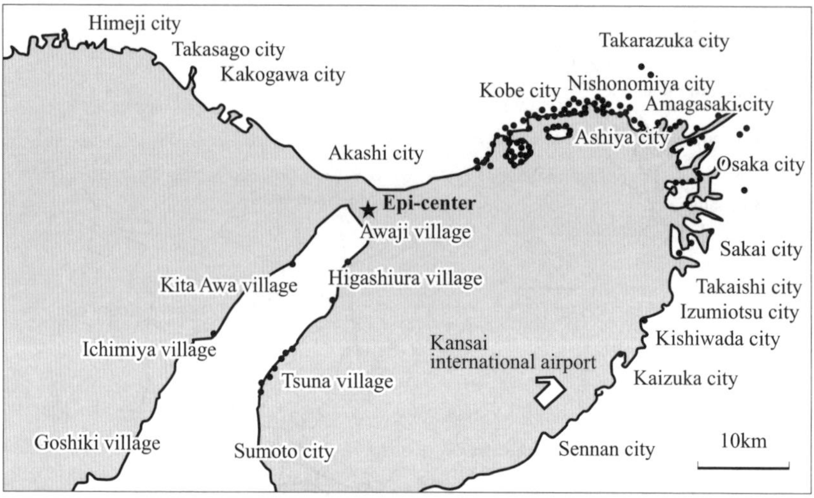


Source: *Shinsai Fukko no Seisaku Kagaku (Policy studies of Post- Earthquake Reconstruction)* Ritsumeikan Daigaku Shinsai Fukko Kenkyu Project (Ritsumeikan University Earthquake Reconstruction Research ProjectÅj (ed), Yuhikaku, p.270 [Japanese].

# *Endnotes*

## Chapter 1

1 This was reported in the media by *Asashi* newspaper, 1998.

2 The original Japanese language version of this book was published in 1994.

3 In 2001, the Ministry of Health and Welfare (MOHW) and the Ministry of Labour (MOL) merged to form the Ministry of Health, Labour and Welfare (MHLW).

4 The figure of 22% is still high by international standards because the social security system developed relatively late and Japan's agricultural and small-business sectors are comparatively large (Ogawa, 1997).

5 Izuhara (2000b: 94–98) has explored this qualitatively with her study in Kita-Kyushu, Japan. Married women are comparatively better off in post-war Japan.

6 *Ikigai* has been translated as life fulfillment (Campbell, 1992: 121). Ingrid Getreuer-Kargl (1995) translated *ikigai* as self-realization and grouped it among the four aspects of Japanese social welfare: economic, health, emotional needs and *ikigai* (Getreuer-Kargl, 1995: 162–168).

7 Izuhara (2000a and b) summarized the changes to the *ie* system in modern Japan over history. The first major change was during the Edo Era when the Tokugawa Shogunate handed over rule to the Meiji Emperor (Izuhara, 2000a: 95, 2000b: 43) The Meiji Civil Code (1898) saw the first changes to the legal structure. The next fundamental change occurred during the US occupation after the Second World War. Ochiai (1994) compared the i*e* system with family and households in Britain, USA, China, and India. The study illustrated the allocation of inheritance to an adopted son or son-in-law, as well as analyse changes in Japan over time. Harada (1996) goes more into legal detail, emphasizing social security.

8 Co-habitation with adult children is also known as co-habitation and co-residence. Izuhara (2000) calls it 'intergenerational living' or 'extended family living'.

9 73.2% across the nation and 56.2% in Tokyo (Miyajima, 1992).

10 This was just before the first international oil crisis in 1973. Until then, Japan had enjoyed a remarkably high rate of economic growth and increasing tax revenues. These generous programs for elderly people

were passed without much difficulty, despite the oil shock, and were pushed forward because the LDP (Liberal Democratic Party) feared a loss of political power (Nishimura, 1993) while the Ministry of Health and Welfare failed to provide an alternative (Campbell, 1992).

11 Imai identified the phenomenon of social hospitalization by highlighting that the average length of stay in Japan is about four times more than the OECD average, reflecting the fact that many acute care beds have taken on the long-term care function for the elderly (Imai, 2002: 7).

12 *Katakana* is one component of the Japanese writing system along with *hiragana* and *kanji*. Most of the katakana scripts are derived from components of more complex *kanji*.

13 *Nagaya* (long-room) housing in Japan is flat housing of only one story which has several small flats for each household. There is a sense of belonging in the streets and among those living in the same *nagaya* unit.

## Chapter 2

1 Crush syndrome is defined as 'Trauma and ischemia of soft tissues, principally skeletal muscle, due to prolonged severe crushing of the tissues, leading to increased permeability of the cell membrane and to the release of potassium, enzymes, and myoglobin from within cells. Ischemic renal dysfunction secondary to hypotension and diminished renal perfusion results in acute tubular necrosis and uraemia' (CancerWEB Project, 2000).

2 Spinal fractures are defined as 'broken bones in the vertebral column' (CancerWEB Project, 2000).

3 Renal replacement therapies are defined as 'Procedures which temporarily or permanently remedy insufficient cleansing of body fluids by the kidneys' (CancerWEB Project, 2000).

4 Acute renal failure (ARF) is defined as 'A sudden decline in renal function [that] may be triggered by a number of acute disease processes. Examples include sepsis (infection), shock, trauma, kidney stones, kidney infection, drug toxicity (aspirin or lithium), poisons or toxins (drug abuse) or after injection with an iodinated contrast dye (adverse effect)' (CancerWEB Project, 2000).

Chronic renal failure represents a slow decline in kidney function over time. It may be caused by a number of disorders which include long-standing hypertension, diabetes, congestive heart failure, lupus or sickle cell anaemia. Both forms of renal failure result in a life-threatening metabolic derangement (CancerWEB Project, 2000). The 1990 Iran earthquake reported a high incidence of ARF (Nadjafi, 1997: 655–64, Eknoyan, 1992: 241–4) as did the 1999 Turkey Marmana and 1988 Armenia Spitak earthquakes (Vanholder et al. 2001: 783–91).

5 A peptic ulcer is defined as 'An ulcer in the wall of the stomach or duodenum resulting from the digestive action of the gastric juice on the mucous membrane when the latter is rendered susceptible to its action' (CancerWEB Project, 2000).

6 The reasons why the high proportion of elderly people reported crush syndrome were more than medical. In a typical Japanese household, elderly people sleep on the ground floor whereas other family members sleep on the upper floor. If an elderly person was sleeping on a futon on a tatami mat on the floor, they would have been more likely to be underneath the fallen furniture at the time of the earthquake.

7 Alcock (1996), Alcock, Erskine and May (2001), George and Miller (1994), Hill (2000), Clasen (1999), Blakemore (1998), Deacon (1997) and Alcock (2000) were checked.

8 The subheadings of *Social Policy* were 'The impact of the mass media upon public attitudes', 'The impact of the media on policy-making', 'Regulation of the press? Satellite, cable, and digital television,' and 'The internet'.

9 During my fieldwork, a welfare commissioner commented that about 90% of residents were in this category, but I was not able to obtain the official figure as it is a sensitive area and no official was willing to share the data with me.

10 Bowling and Browne (1991) further describe the importance of community networks for health in old age.

## Chapter 3

1 The fieldwork sites visited over the course of this study are represented here with pseudonyms. As the reports which provided the statistical background for the study reveal the actual names of these sites, they cannot be cited in this publication.

2 Japan's total population is half that of the USA. The largest US papers, *USA Today*, the *Wall Street Journal*, the *New York Times* and the *Los Angeles Times* have circulations between 1 and 2 million as of late 1994 (Cooper-Chen, 1997: 52).

3 NVivo is computer assisted qualitative data analysis software (Richards, 1999a, 1999b, 1999c). It is widely used, especially in Australia, the UK, and South Africa. It is also becoming popular in the USA. NVivo was released in May 1999 as a more updated version of Nudist 4 (Fraser, 1999). I received training on NVivo (di Gregorio, 1999) in London, October 1999 and attended a week-long workshop on research methods in Australia in November 1999 where the Nudist & NVivo developers, Prof. Lyn Richards and Dr. Pat Bazeley were lecturers. My field data was used as workshop materials for demonstration and as an exercise. This research is among the first to applied NVivo in a non-English context.
4 The rearranged version of this section has been published online at the Sage homepage in November 2009 as Project 1: Elderly survivors of the 1995 Great Hanshin Earthquake in Japan (Otani, 2009c).
5 Unlike, for example, American English, Japanese is a high-context language. In American English you have to spell everything out; you cannot expect people to read between the lines. I learned this while working at the World Bank. The World Bank is an international organization based in Washington DC working for poverty alleviation in the developing world. While the World Bank uses American English because it is a low-context language, most clients of the World Bank, actually 80% of the world population, use high-context languages. It is easier for those with high context languages to move to low context, but it is more difficult to move from low context to high context.

## Chapter 4

1 The fiscal year in Japan starts in April and ends in March.
2 The GHQ test is a self-report screening test designed to detect psychiatric disorders among outpatients and those in the general community. It focuses on two areas: the inability to carry out normal functions and the appearance of new distressing phenomena. GHQ 30 is a short form of the GHQ 60, while the GHQ 28 provides more scores, including severity, somatic symptoms, anxiety, insomnia, social dysfunction and severe depression. It takes 3–4 minutes to administer.
3 IES-R was developed in 1997 to test and score post-traumatic stress disorder (PTSD).
4 The results revealed that elderly respondents reported fewer threat and disruption experiences and used fewer general and disaster-related support services. However, older adults reported higher overall levels

of post-traumatic stress symptoms on the IES than younger adults (Tichehurst et al., 1996).

## Chapter 6

1 Since the fiscal year of 1987, the Silver Housing program has been undertaken with liaison between the housing and welfare services. The program involves the provision of public housing with services such as emergency aid, care during illness, counselling and guidance on daily life. In addition to LSAs, welfare facilities are either attached or located adjacent to or in the neighborhood of Silver Housing (Enomoto, 1996).

2 The LSA office has an emergency system. A resident of an apartment for elderly people can make an emergency call from their room. If a resident does not use water for more than 12 hours, an alarm is activated and the LSA goes to check on them. The system is also linked to an ambulance.

3 *Dōwa* means sameness or harmony. It is the movement to end discrimination against *burakumin*, the outcasts. They were the untouchable lowest castes in the class system during the Edo Era. The notion was legally abolished several hundred years ago but discrimination continues; *burakumin* tend to live in the same areas, do not receive high educations and generally engage in occupations such as butchers, shoe-makers, and rubbish collectors.

4 See Enomoto (1996) for further details.

5 Japan enjoys the longest average life expectancy in the world, but this varies among regions. According to Cockerham and Yamori (2001), variations occur due to the practice of summer festival, as well as malnutrition and unhealthy habits.

6 *Minsei iin* are statutory voluntary workers. The Welfare Commissioner Law of 1948 stipulates that *minsei iin*, designated by the minister of health and welfare, monitor the needy in each town district. *Minsei iin* do not receive a salary and are reimbursed only for expenses, but they carry prestige and exercise discretionary power in the community (Hashimoto, 1996: 45).

7 While PRH is still a public issue in Kobe, it is not in Tokyo. As time progresses, it is more difficult to gain attention and financial support from the central government. One incident that attracted attention away from the Hanshin earthquake was the March 1995 Aum terrorist attack in the Tokyo subway, three months after the earthquake. In Kobe, job

creation has been the most pressing issue as the economy has been in decline since the earthquake.

## Chapter 7

1 *Seikatsu fukkō* (life reconstruction) often featured in Kobe media broadcasts during my fieldwork. However, I was surprised to find that in Tokyo the word was uncommon. It was one of the new words which began to be used after the earthquake.
2 Enomoto (1996) was first published in Japanese in 1994.
3 This study was derived from PhD research conducted in 1969.
4 In my grandmother's generation, married woman could not return home other than for a New Year greeting. The Japanese white wedding kimono dress symbolized a death and rebirth into a new family, not virginity, as it does in Western culture.
5 Japan Overseas Christian Medical Cooperative Services (JOCS) is a NGO sending medical staff to developing countries, mainly in Asia.
6 Campbell translated *ikigai* as life fulfillment (Campbell, 1992: 121). Ingrid Getreuer-Kargl (1995) translated *ikigai* as self-realization. She concluded that social welfare in Japan needed to provide for economic welfare, health, emotional welfare and the need for self-realization. The Ministry of Health and Welfare White Paper stated that, 'Individuals themselves must prepare for old age; namely take care to preserve their health, set up relationships and ensure a meaning in life' (in Getreuer-Kargl, 1995: 162–168). Here, *ikigai* is translated as 'meaning in life'.

## Chapter 8

1 This data appeared in an article titled, '*Kodokushi* and suicides. 38 people in the past one year'.
2 The author did not use the Japanese term, '*kodokushi*'. Only the English translation was used in this English journal article.

# References


Alcock, C. ( 2000 ) *Introducing Social Policy*, Prentice Hall, Pearson Education Limited, Essex.

Alcock, P (1996) *Social Policy in Britain: Themes & Issues*. Macmillan Press.

Alcock, P, Erskine, A, and May, M. (eds.) (2001) *The student's companion to social policy*, Blackwell.

Alpass, F.M. and Neville, S. (2003) 'Loneliness, health and depression in older males'. *Aging and Mental Health*. 7(3): 212–216.

Altheide, D.L. (1996) *Qualitative Media Analysis*. Qualitative Research Methods Series. Maanen, J.V. (ed.)Vol.38. Thousand Oaks, Sage.

Amella, E. (2002) 'Resistance at mealtimes for persons with dementia'. *Journal of Nutrition, Health and Aging*, 6(2):117–22.

Anastassiadou, V. and Heath, M. (2002) 'Food choices and eating difficulty among elderly edentate patients in Greece'. *Gerontology*,19(1):17–24.

Antaki, C. (ed.) (1988) *Analysing everyday explanation: a casebook of methods*. London: Sage.

Arber, S. (1993) 'Designing samples', in Gilbert, N. (ed.), *Researching Social Life*, London: Sage Publications. pp.68–93.

Arber, S. and Ginn, J. eds, (1995) *Connecting gender and ageing: a sociological approach*. Buckingham: Open University Press:

Arber, S. and Cooper, H. (1999) 'Gender differences in health in later life: the new paradox?' *Social Science and Medicine*. 48: 61–76.

Armenian, H., et al. (2000) 'Loss as a determinant of PTSD in a cohort of adult survivors of the 1998 earthquake in Armenia: implications for policy'. *Acta Psychiatr Scand*, 102(1): 58–64.

Armenian, H.K., et al. (1997) 'Deaths and Injuries due to the Earthquake in Armenia: A Cohort Approach'. *International Journal of Epidemiology*, 28(4): 806–813.

Armenian, H.K., Melkonian, A.K. and Hovanesian, A.P. (1995) 'Long term mortality and morbidity related to degree of damage following the 1988 earthquake in Armenia'. *American Journal of Epidemiology*, 148(11): 1077–84.

Armenian, H.K., Noji, E.K. and Oganesian, A. (1992) 'A case-control study of injuries arising from the earthquake in Armenia, 1988'. *Bulletin of the World Health Organization*, .70(2): 251–257.

Ashita no nihon wo tsukuru kyokai (2001) *Report on the self-governing body support system for older people.* Online: http://www.ashita.or.jp/shiryou/jichikai. Accessed 14 Nov 2003.

Atkins, D. Moy, E.M. (2005) 'Left behind: the legacy of Hurricane Katrina: Hurricane Katrina puts the health effects of poverty and race in plain view', *British Medical Journal,* 331, 22: 916–918.

Azarian, A.et al. (1996) 'Baseline assessment of children traumatized by the Armenian Earthquake'. *Child Psychiatry Hum Dev*, 27(1): 29–41.

Baldock, J. (ed.) (1994) *Social Policy.* New York: Oxford University Press.

Barusch, A.S. (1994) *Older Women in Poverty: Private Lives and Public Policies.* New York Springer Publishing Company.

Bass, S.A. and Oka, M. (1995) An older worker employment model: Japan's Silver Human Resource Centres, *Gerontologist*, 35(5) 679–82.

Bauer, M.W. and Gaskelleds, G. (2000) *Qualitative researching with text, image and sound: A practical handbook.* London: Sage Publications.

Bazeley, P. and Richards, L. (2000) *The NVivo Qualitative Project Book.* London: Sage Publications.

BBR and International Partners (2005) *One year after the Tsumanmi: The recovery effort and way forward.* December.

Becker, H.S. and Richards, P. (1986) *Writing for social scientists: How to start and finish your thesis, book, or article.* Chicago and London: University of Chicago Press.

Bender, E. (2005) 'Psychiatrists Find Tsunami MH Consequences Severe: International News', *Psychiatric News,* August 5, 40 (15)

Bentelspacher, C. and Minaieds, K. (1994) *Ageing in Japan and Singapore.* Singapore: Department of Japanese Studies, National University of Singapore.

Bhalla, N. (2006) Special Article: Relief after the wave. *British Dental Journal*, 200(2): 116–118.

Biggs, S. (1993) *Understanding ageing: images, attitudes & professional practice.* Buckingham: Open University Press.

Blakemore, K. (1998)*Social Policy An Introduction*, Open University Press, Buckingham,

Bland, S.H., et al. (1996) 'Long-term psychological effects of natural disasters'. *Psychosomatic Medicine.* 58:18–24.

Bland, S.H., et al. (1997) 'Social network disturbances and psychological distress following earthquake evacuation'. *Journal of Nervous and Mental Disease.*185(3): 188–194.

Boling, P. (1998) 'Family Policy in Japan'. *Journal of Social Policy.* 27 (2): 173–190.

Bowling, A. (1991a) 'Social support and social networks: their relationship to the successful and unsuccessful survival of elderly people in the community. An analysis of concepts and a review of the evidence'. *Family Practice*, 8(1): 68–83.

Bowling, A. (1991b) *Measuring health: a review of quality of life measurement scales*. Milton Keynes: Open University Press. 20–32.

Bowling, A. and Browne, P.D. (1991) 'Social network, health, and emotional well-being among the oldest old in London'. *Journal of Gerontology*, 46(1): S20–32.

Bowling, A., E. Grundy, and Farquhar, M. (1997) *Living well into old age: three studies of health and well-being among older people in East London and Essex*. London: Age Concern England.

Bradbury, M. (1999) *Representations of death: A social psychological perspective*. London: Routledge.

Brink, S. (1998) *Housing older people: an international perspective*. New Brunswick, USA and London, UK: Transaction Publishers.

Bryman, A. and Burgess R.G. (eds) (1994) *Analyzing Qualitative Data*. London: Routledge.

Burke, D., et al. (1993) The development of a spinal injuries unit in Armenia. *Paraplegia*, 31(3): 168–71.

Burman, E. and Parkereds, I. (1993) *Discourse analytic research: Repertoires and readings of texts in action*. London: Routledge.

Burrows, K. (1997). *Meeting the needs of the elderly in natural disaster*. Eastern Nebraska Office on Aging, Omaha, NE.

Butterfoss, F., et al. (1993) 'Community coalition for health promotion and disease prevention'. *Health Education Research*.8(3): 315–330.

Byleveld, PM. Kent, MI and McCall, BJ (2005) Operation Sumatra Assist: post-tsunami environmental and public health response in Banda Aceh, Australian Defence Force Health, 6: 48–53.

Campbell, J.C. (1992) *How policies change: The Japanese government and the aging society*. Princeton, New Jersey: Princeton University Press.

Campbell, J.C. (1996) 'Media and policy change in Japan', in Pharr, S.J. and Krauss, E.S. (eds.) *Media and Politics in Japan*, Honolulu, Hawaii: University of Hawaii Press: 187–212.

Campbell, J.C. and Ikegami, N. (1998) *The Art of Balance in Health Policy: Maintaining Japan's Low-Cost, Egalitarian System*. New York: Cambridge University Press.

Campbell, J.C. and Ikegami, N. (2000) 'Long-term-Care Insurance comes to Japan'. *Health Affairs*,19(3): 26–39.

Cancer Web Project (2000) *CancerWeb Online Medical Dictionary* Online: http://cancerweb.ncl.ac.uk/omd.

Cardena, E. and Spiegel, D. (1993) 'Dissociative reactions to the San Francisco Bay Area earthquake of 1989'. *American Journal of Psychiatry*, 150(3): 474–8.

Carr, V.J., et al. (1997a) 'Psychosocial sequelae of the 1989 Newcastle earthquake: II. Exposure and morbidity profiles during the first 2 years post-disaster'. *Psychosomatic Medicine*.27: 167–178.

Carr, V.J., et al. (1997b) 'Psychosocial sequelae of the 1989 Newcastle earthquake: III. Role of vulnerability factors in post-disaster morbidity'. *Psychosomatic Medicine*, 27: 179–190.

Carr, V.J., et al. (1997c) 'A synthesis of the findings from the quake impact study: a two-year investigation of the psychosocial sequelae of the 1989 Newcastle earthquake. *Soc Psychiatry Psychiatr Epidemiol*.32: 123–136'.

Centers for Disease Control (US CDC) (2005)'Assessment of health-related needs after Tsunami and Earthquake, three districts, Aceh Province, Indonesia, July–August 2005, *MMWR(CDC Morbidity and Mortality Weekly Report)* February 3, 2006 / 55(04);93–97

Centers for Disease Control (US CDC) (2006) 'Assessment of health-related needs after Tsunami and Earthquake, three districts, Aceh Province, Indonesia, July–August 2005, *MMWR (CDC Morbidity and Mortality Weekly Report)* February 3, 2006 / 55(04);93–97

Chan, E.Y.Y. (2008) 'The untold stories of the Sichuan earthquake', *Lancet* 372: 359–362

Chanan, G. (1992) *Out of the shadows: Local community action in the European community,* Dublin: European Foundation for the Improvement of Living and Working Conditions.

Chattergee, P. (2005) 'Mental health care for India's tsunami survivors', *Lancet,* 365: 5: 833–834.

Cheng, MH. (2006) Post-tsunami boost to southeast Asia's mental health care: world report, *Lancet*, 7: 367: 15–17.

Cirillo, L. (1994) 'Verbal imaginery of aging in the news magazines', in Shenk, D. and Achenbaum, W.A. (eds.) *Changing perceptions of Aging and the Aged.* New York: Spinger Publishing Company. 173–178.

Clasen, J. (ed.) (1999) *Comparative social policy: concepts, theories and methods*. Blackwell.

Cockerham, W.C. and Yamori, Y. (2001) 'Okinawa: an exception to the social gradient of life expectancy in Japan'. *Asia Pacific Journal of Clinical Nutrition*. 10(2): 154–158.

Connell, R.W. (1995) *Masculinities.* Berkeley and Los Angeles: University of California Press.

Cooper-Chen, A. (1997) *Mass Communication in Japan.* Ams, Iowa: Iowa State University Press.

Creswell, J.W. (1994) *Research Design: Qualitative & Quantitative Approaches.* Thousand Oaks: Sage.

Creswell, J.W. (2003) *Research Design: Qualitative, Quantitative and Mixed Methods Approaches*, Second edition, Thousand Oaks: Sage.

de Castro, J. (2002) 'Age-related changes in the social, psychological, and temporal influences on food intake in free-living, healthy, adult humans'. *Journal of Gerontology A Biol Sci Med Sci.* 57(6): M368–77.

Deacon, B (1997) *Global Social Policy: International organization and the future of welfare.* Sage.

de Jong, D. S. Prosser, N. Ford, N. (2005) 'Addressing psychosocial needs in the aftermath of the tsunami'. *PLoS Medicine* 2(6): 179.

di Gregorio, S. (1999) *Introduction to QSR NVIVO workshop.* London.

Dominelli, L. (1990) *Women and Community Action*, Birmingham: Venture Press.

Doshisha University (1997) *Isolated deaths: miniature of welfare state without guarantee for life: report of survey of life and health of the middle-aged in temorary shelters.* Kyoto: Doshisha Daigaku.

Drake, R.F. (1998) 'Housing and older people', in A. Symonds and A. Kelly (eds.) *The Social Construction of Community Care*, Houndmills and London: Macmillan Press Ltd. 254 261.

Eknoyan, G. (1992) 'Acute renal failure in the Armenian earthquake'. *Renal Failure.*14(3): 241–4.

Ely, M., et al. (1997) *On Writing Qualitative Research: Living by Words.* London: Falmer Press.

Emerson, R. (1995) *Writing ethnographic field notes.* University of Chicago Press.

Enomoto, K. (1996) *Houses and Welfare Services for the Aged: Link Services of Welfare and Collective Housing for the Aged in Iceland and Japan.* Seizansha: Sagamihara.

Falk, H. G. Baldwin, G. (2006) 'Environmental Health and Hurricane Katrina', *Environmental Health Perspectives*, 114(1): A12–13.

Fanelli, M. and K. Stevenhagen, K. (1985) 'Characterizing consumption patterns by food frequency methods: core foods and variety of foods in diets of older Americans'. *Journal of American Diet Association*, 85(12): 1570–6.

FBI (2009) Uniform crime reports statistics. Online: http://www.fbi.gov/ucr/cius2008/index.html.

Fielding, N.G. and R.M. Lee (1991) *Using Computers in Qualitative Research.* London: Sage.

Fischer, C. and S. Phillips (1982) 'Who is alone? Social characteristics of people with small networks', in L.A. Peplau and D. Perlman (eds) *Loneliness: A sourcebook on current theory, research, and therapy,* New York: Wiley.

Flick, U. (1998) *An Introduction to Qualitative Research.* London: Sage.

Fletcher, LE, E. Stover and H.M. Weinstein, HM (eds.) (2005) *After the Tsunami: Human Rights of Vulnerable Populations.* Berkeley: Human Rights Center, University of California, Berkeley, Honolulu: East-West Center.

Florin, P., R. Mitchell, and J. Stevenson, (1993) 'Identifying training and technical assistance needs in community coalitions: a developmental approach'. *Health Education Research,* 8(3): 417–432.

Forrest, R., A.L. Grange and N.M. Yip (2002) 'Neighbourhood in a high rise, high density city: some observations on contemporary Hong Kong'. *The Sociological Review,* 50(2): 215–240.

Franklin, B. (ed.) (1999) *Social policy: the media and misrepresentation.* London: Routledge.

Fraser, D. (1999) *QSR Nudist Vivo Reference Guide.* Melbourne: Qualitative Solutions and Research Pty. Ltd.

Fujii, Y. (1993) 'Pension and independence of women', in *Women and Social Security* Research Institute of Social Security: University of Tokyo Press.

Fukushi Channel (2002) *Annual Report on establishment of the society of co-participation of men and women, 2001*, Online: http://www.fukushi.com/news/2002/07/020701-a.html, Accessed 14 November 2003.

George, V and S. Miller (eds.) (1994) *Social Policy towards 2000: Squaring the Welfare circle*, London and New York, Routledge.

Gerrity, E.T. and B.W. Flynn (1997) 'Mental health consequences of disasters', in E.K Noji (ed.) *The Public Health Consequences of Disasters,* New York and Oxford: Oxford University Press. 101–121.

Getreuer-Kargl, I. (1995) 'Social policy planning and women in Japan', in B. Kosiaho (ed.) *Women, the Elderly and Social Policy'* , Ashgate: Aldershot:147–172.

Gibson, H. (2000) *Loneliness in later life.* Hampshire and London: Macmillan Press.

Giddens, A. (1990) *The Consequences of Modernity.* Cambridge: Polity.

Giddens, A. (1991) *Modernity and Self-Identity: Self and Society in the Late Modern Age*. Cambridge: Polity.

Gill, R. (1996) 'Discourse analysis: practical implementation', in J.T.E. Richardson (ed.) *Handbook of qualitative research methods for psychology and the social sciences*. Leicester: BPS Books 141–158.

Gill, R. (2000) 'Discourse Analysis', in M.W. Bauer and G. Gaskell (eds.) *Qualitative researching with text, image and sound: a practical handbook,* London, Thousand Oaks, New Delhi: Sage, 172–190.

Gilloran, J. (1968) 'Social problems associated with "high living"'. *Medical Officer*, 120: 117–118.

Gleser, GL. Green, BL. Winget, C. (1981) *Prolonged psychosocial effects of disaster*. New York: Academic Press.

Goenjian, A. (1993) 'A mental health relief programme in Armenia after the 1988 Earthquake: Implementation and clinical observations'. *British Journal of Psychiatry*,163: 230–239.

Goenjian, A.K., et al. (1994a) 'Posttraumatic Stress Disorder in Elderly and Younger Adults after the 1988 Earthquake in Armenia'. *American Journal of Psychiatry*, 151(6): 895–901.

Goenjian, A.K., et al. (1994b) 'Posttraumatic stress reactions after single and double trauma'. *Acta Psychiatr Scand*. 90: 214–221.

Goenjian, A., et al. (1995) 'Psychiatric comorbidity in children after the 1998 earthquake in Armenia'. *Journal of American Academic Child and Adolescent Psychiatry*, 34(9): 1174–84.

Goenjian A.K., Walling, D., Steinberg, A.M., Karayan, I., Najarian, L.M., Pynoos, R. (2005) 'A prospective study of posttraumatic stress and depressive reactions among treated and untreated adolescents 5 years after a catastrophic disaster'. *American Journal of Psychiatry*. 162(12):2302–8.

Goenjian, A., et al. (1997) 'Outcome of psychotherapy among early adolescents after trauma'. *American Journal of Psychiatry*, 154(4): 536–42.

Goldstein, E., et al. (1996) *Trends in health status, services, and finance: The transition in Central and Eastern Europe, Volume I*. Washington, D.C.: The World Bank.

Gordon, M. (1996) 'What makes life worth living?: How Japanese and Americans make sense of their words', Berkeley: University of California Press.

Gubrium, J.F. (1988) 'Analyzing field reality'. *Qualitative research methods series*. Manning, P.K. Maanen, J.V. and Miller, M.L. (eds.) Vol.8. Newbury Park, CA: Sage.

Gurley, R. J. Lum, N. Sande, M. Lo, B. and Katz, M.H. (1996) 'Persons found in their homes helpless or dead'. *New England Journal of Medicine*. 334 (26): 1710–1726.

Gustafsson, K. and Sidenvall, S. (2002) 'Food-related health perceptions and food habits among older women'. *Journal of Advanced Nursing*, 39(2): 164–73.

Hagemann, R.P. and Nicoletti, G. (1990) *Population Ageing: Economic effects and some policy implications for financing public pensions*. 51–78.

Haines, V.A., Hurlbert, J.S. and Beggs, J.J. (1996) 'Exploring the determinants of support provision: Provider characteristics, personal networks, community contexts, and support following life events'. *Journal of Health and Social Behavior,* 37: 252–264.

Hanson, P. (1989) 'Citizen involvement in community health promotion: a role application of CDC's PATCH model'. *International Quarterly of Community Health Education*, 9(3): 177–186.

Harada, S. (1996) *The Aging Society, the Family, and Social Policy.* Tokyo: University of Tokyo.

Harre, R. (1998) 'Accountability: the role of pronouns', in Antaki, C. (ed.) *Analysing Everday Explanation: A casebook of Methods'*, London: Sage Publications. 156–167.

Hashimoto, A. (1996) *The Gift of Generations: Japanese and American Perspectives on Aging and the Social Contract,* Cambridge University Press

Hashimoto, R. (1995) '*Minsei iin* community volunteers: an important network in the field of welfare in Japan', in Koskiasho, B. (ed.) *Women, the elderly and social policy in Finland and Japan,* Avebury: Aldershot. 190–196.

Hayashi, F. (1992) 'Explaining Japan's saving: A review of recent literature'. *Bank of Japan Monetary and Economic Studies*.10(2): 63–78.

HelpAge International (HAI) (2005a) *Life after the tsunami, Ageing and Development*. 17 Feb.

HelpAge International (HAI) (2005b) The impact of the Indian Ocean tsunami on older people: Issues and recommendations.

Heywood, F., Oldman, C. and Means, R. (2002) *Housing and home in later life*, Buckingham, Philadelphia: Open University Press.

Hill, M. (2000) *Understanding Social Policy, 6th Edition*. Blackwell,

Hirayama, Y. (2000) 'Collapse and reconstruction: Housing recovery policy in Kobe after the Hanshin Great Earthquake'. *Housing Studies*.15(1): 111–128.

Hiroi, Y. (1998) *Social Security and Economic Development: Japanese Experience*. Bangkok.

Hiroshima, K. (1992) 'The living arrangements and family contracts of the elderly in Japan',. in: Kreishnan, P. and Mahadevan, K. (eds) *The elderly population in developed and developing world: policies, problems and perspectives*. Delhi: B.R. Publishing Corporation. 68–86

Hoch, C. and Slayton, R. (1989). *New homeless and old: Community and the skid row hotel*. Philadelphia: Temple University Press.

Hochschild, A.R. (1973) *The Unexpected Community*. Englewood Cliffs, New Jersey: Prentice-Hall, Inc.

Hollway, W. and Jefferson, T. (2000) *Doing qualitative research differently: free association, narrative and the interview method*. London: Sage Publications.

Holmen, K. (2002) Loneliness, health and social network among elderly people—a follow up study, *Archives of Gerontology and Geriatrics*, 35: 261–274.

Hoshino, S. (1996) 'Paying for the health and social care of the elderly', in Bass, S.A. Morris, R. and Oka, M. (eds.) *Public policy and the old age revolution in Japan*, Binghamton, NY: Haworth Press. 37–16.

Hughes, B. (1995) *Older people and community care: critical theory and practice*. Buckingham: Open University Press.

Hyogo Prefecture Nursing College) (1996) [Japanese] *Community generation at the second Murotani temporary shelters*, Kobe City West Ward health office.

Ikegami, N. and Hasegawa, T. (1995) 'The Japanese health care system: A stepwise approach to universal coverage', in Dunlop, D.W. and Martins, J.M. (eds.) *An international assessment of health care financing: Lessons for developing countries*? Washington D.C.: Economic Development Institute of the World Bank. 33–64.

Ikegami, N. and Campbell, J.C. (1995) 'Medical Care in Japan'. *New England Journal of Medicine*, 333(19): 1295–1299.

Ikegami, N. and J.C. Campbell (eds.) (1996) *Containing health care costs in Japan*. Ann Arber: The University of Michigan Press.

Iliffe, S., et al. (1992) 'Are elderly people living alone an at risk group?' *British Medical Journal*, 305: 1001–1004.

Illiffe, S. (1997) 'Old, frail people are marginalised in Japan', *British Medical Journal*. 519

Imai (2002) *Health Care Reform in Japan*. Paris: OECD

Iwasaki, N., Ukai, K., Urano, M., Tsuji, K., Nitagai, K., Noda, T., and Yamamoto, T. (eds.) (1999) [Japanese] *Sociology of the Great Hanshin Earthquake: Sociology of Evacuee Life*. Kyoto: Shōwadō.

Izuhara, M. (2000a) *Family change and housing in post-war Japanese society: The Experiences of older women*. Ashgate.

Izuhara, M. (2000b) 'Changing family tradition: Housing choices and constraints for older people in Japan'. *Housing Studies*, 15(1): 89–110.

Jerrome, D. (1992) *Good Company: An Anthropological Study of Old People in Groups*. Edinburgh: Edinburgh University Press.

Jerrome, D. (1999) 'Stability and change in late-life friendships', *Ageing and Society* 19(6): 661–676.

Kano, M. (2000) 'The impact of modernisation and social policy on family care for older people in Japan'. *Journal of Social Policy*, 29(2): 181–203.

Karanci, A.N. and Rustemli, A. (1995) 'Psychological consequences of the 1992 Erzincan (Turkey) Earthquake'. *Disasters,* 19(1): 8–18.

Kario, K., et al. (1997a) 'Earthquake-induced potentiation of acute risk factors in hypertensive elderly patients: Possible triggering of cardiovascular events after a major earthquake. *Journal of American College of Cardiology*, 29(5): 926–33.

Kario, K. and Ohashi, T. (1997b) Increased coronary heart disease mortality after the Hanshin-Awaji Earthquake among the older community of Awaji Island. *Journal of American Geriatric*s Society, 45(3): 610–613.

Kato, H., et al. (1996) 'Post-traumatic symptoms among younger and elderly evacuees in the early stages following the 1995 Hanshin-Awaji earthquake in Japan'. *Acta Psychiatr Scand*, 93(6): 477–81.

Keith, J (1982) Old People, New Lives: Community Creation in a Retirement Residence. Chicago: University of Chicago Press.

Kendig, H.L., Hashimoto, A. and Coppard, L.C. (eds) (1991) *Family Support for the elderly: the international experience*, Oxford: Oxford University Press.

Kinoshita, Y. and Kiefer, C.W. (1992) *Refuge of the honored: Social organisation in a Japanese retirement community*. Berkeley and Los Angeles, California: University of California Press.

Kleinman, A. (1995) *Writing at the Margin: Discourse between anthropology and medicine*. Berkeley and Los Angeles, California: University of California Press.

Klinenberg, E. (1999) Denaturalizing disaster: A social autopsy of the 1995 Chicago heat wave. *Theory and Society* 28: 239–295.

Klinenberg, E. (2002) *Heat wave: a social autopsy of disaster in Chicago*. Chicago: University of Chicago Press.

Knight, B.G. Gatz, M. Heller, K. and Bengtson, V.L. (2000) 'Age and emotional response to the Northridge Earthquake: A longitudinal analysis'. *Psychology andAging*. 15 (4) 627–634.

Kobe City West Ward Health office (1998) *Community generation at temporary shelters and Seishinminami public reconstruction housing.*

Kobe University (1997a) *Kutou no Hisai Seikatsu.*Hanshin Daishinsai Kenkyu, ed. Vol.2. Kobe: Kobe Shinbun Shuppan Centre.

Kobe University (1997b) *Kobe no Fukkou wo motomete.*Hanshin Daishinsai Kenkyu ed. Vol.3. Kobe: Kobe Shinbun Sougou Shuppan Centre.

Kose, S. (1998) 'Housing Elderly People in Japan', in Brink, S. (ed.) *Housing Older People: an international perspective*, New Brunswick: Transaction Publishers. p.125–139.

Kostelny, K. and Wessells, M. (2005) 'Psychosocial aid to children after the Dec 26 tsunami', *Lancet,* 366: 2066–7.

Koyana, W. (1996) 'Filial piety and intergenerational solidarity in Japan', *Australian Journal of Ageing*, 15(2): 51–6.

Krause, N. (1987). Exploring the impact of a natural disaster on the health and psychological well-being of older people. *Journal of Human Stress.* Summer: 61–69.

Krieger, J. and Higgins, D.L. (2002) 'Housing and Health: Time again for public health action'. *American Journal of Public Health*, 92(5): 758–768.

Kuwagata, Y., et al. (1997) 'Analysis of 2,702 Traumatized Patients in the 1995 Hanshin-Awaji Earthquake.' *Journal of Trauma: Injury, Infection, and Critical Care,* 43(3): 427–432.

Liddiard, M. (1999) 'Arts and Cultural Policy', in Baldock et al.(ed.) *Social Policy, 2nd edition.* Oxford University Press.

Liddiard, M. (2007) 'Welfare, Media and Culture', in Baldock et al. (ed.) Social Policy, 3rd edition. Oxford University Press.

Lima, B., et al. (1992a) 'Psychiatric disorders among victims of disasters in Ecuador'. *Bol Oficina Sanit Panam,* 113(1): 28–34.

Lima, B., et al. (1992b) 'Psychiatric disorders among emotionally distressed disaster victims attending primary mental health clinics in Eduador'. *Bulletin of the Pan American Health Organization,* 26(1): 60–6.

Lima, B., et al. (1993) 'Emotional distress in disaster victims: A follow-up study'. *Journal of Nervous and Mental Disease,* 181(6): 388–93.

Lock, M. (1996) Displacing Suffering: The Reconstruction of Death in North America and Japan'. *Deadalus*, 125: 207–44.

Long, S.O. and Harris, P.B. (2000) 'Gender and Elder Care: Social change and the role of caregiver in Japan'. *Social Science Japan Journal*, 3(1): 21–36.

Louis, M., et al.(1996) 'Relocation after a disaster: posttraumatic stress disorder in Armenia after the earthquake'. *Journal of American Academic Child and Adolescent Psychiatry,* 35(3): 374–83.

MacIntosh, C. and Morley, J. (2001) 'Effect of exogenous cholecystokinin

(CCK)-8 on food intake and plasma CCK, leptin, and insulin concentrations in older and young adults: Evidence for increased CCK activity as a cause of the anorexia of aging'. *Journal of Clinical Endocrinology and Metabolism* 86. 5830–5837.

Maeda, D. (1989) 'Decline of family care and the development of public services: A sociological analysis of the Japanese experience', in: Eekelaar, J.M. and Pearl, D. (eds.) *An aging world: Dilemmas and challenges for law and social policy.* Oxford: Clarendon Press.

Marshall, T., et al. (2002) 'Oral health, nutrient intake and dietary quality in the very old'. *Journal of American Dental Association*, 133(10): 1369–79.

Maruo, S. and Matumoto, M. (1996) 'Spinal fractures resulting from the 1995 Great Hanshin Earthquake of the Kobe-Osaka area of Japan', *Spinal Cord*, 34; 382–386.

Massey, B.A. (1997) 'Victims or survivors? A three-part approach to working with older adults in disaster'. *Journal of Geriatric Psychiatry.* 30 (1): 193–202.

Matsuzuka, Y. (2002) *Changes in the permanent employment system in Japan: Between 1982 and 1997.* New York and London: Routledge.

Mattox, KL. (2006) 'Hurricanes Katrina and Rita: Role of individuals and collaborative networks in mobilizing/coordinating societal and professional resources for major disasters', *Critical Care*, 10:205.

Maykovich, M. (1979) 'The Japanese family', in Das, M. and Bardis, P. (eds.) *The Family in Asia*', New Delhi: Vikas Publishing House. 381–410.

McDonnell, S., et al. (1995a) 'Evaluation of long-term community recovery from Hurricane Andrew: Sources of assistance received by population sub-groups'. *Disasters*, 19(4): 338–347.

McDonnell, S., et al. (1995b) 'Long-term effects of Hurricane Andrew: Revisiting mental health indicators', *Disasters* 19(3): 1995.

Mecocci, P. Di, I.A. Pezzuto, S. Rinaldi, P. Simonelli, G. Maggio, D. Montesperelli, P. Longo, A. Chrubini, A. and Chiarappa, N. (2000) 'Impact of the earthquake of September 26 1997 in Umbria, Italy on the socioenvironmental and psychophysical conditions of an elderly population'. *Aging: Clinical and Experimental Research.* 12 (4): 281–286.

Melkonian, A.K., Armenian H.K., and Hovanesian, A.P.(1997) 'Long-term studies of mortality and morbidity following the Armenian earthquake, Earthquakes and People's Health: Vulnerability reduction, Preparedness, Rehabilitation', *Proceedings of a WHO symposium, Kobe,*

*27–30 January 1997.* World Health Organization Centre for Health Development: Kobe.

Mellor, P.A. (1993) 'Death in high modernity: the contemporary presence and absence of death', in Clark, D. (ed.) *The Sociology of Death: Theory, culture, practice*, Oxford and Cambridge: Blackwell Publishers.

Miller, T., et al. (1993) 'Post-traumatic stress disorder in children and adolescents of the Armenian earthquake', *Child Psychiatry Human Development*, 24(2): 115–23.

Miyajima, H. (1992) [Japanese] *Socio-Economics of Ageing Society*. Iwanami Shoten: Tokyo.

Miyajima, H. (1992) 'The family structure in contemporary Japan', in Miyajima, H. (ed.) *Socioeconomics for the Era of Aging: the Family, the Firm, and the Goverment*, Iwanami Shoten: Tokyo: 77–104.

MOHW (1997) *White paper: Annual report of Health and Welfare*. Tokyo: MOHW

MOHW (1996) *The Japanese experience in social security*. The East Asian Ministrial Meeting on Caring Societies.

Morioka, H., Nagatomo, I. and Takigawa, M (1996) *Alcohol consumption patters of medical students measured by the Kurihama alcoholism screening tests*. Kagoshima.

Morrow, B.H. (1999) 'Identifying and mapping community vulnerability', *Disasters*, 23(1): 1–18.

Moser, C.O.N. (1993) *Gender planning and development: Theory, practice and training*. London: Routledge.

Moser, C. and McIlwaine, C. (1997) *Household Responses to Poverty and Vulnerability Volume 3: Confronting Crisis in Commonwealth, Metro Manila, the Philippines*. Washington, D.C.: World Bank.

Moser, C.O.N. and Peakceds, L. (1987) *Women Human Settlements and Housing*. London: Tavistock Publications.

Nadjafi, I., M. R. Atef, B. Broumand and A. Rastegar (1997) 'Suggested Guidelines for Treatment of Acute Renal Failure in Earthquake Victims', *Renal Failure*, 19(5): 655–664.

Najarian, L.et al. (1996) 'Relocation after a disaster: posttraumatic stress disorder in Armenia after the earthquake', *Journal of American Academy of Child Adolescent Psychiatry*, Mar; 35(3): 374–83.

Najarian, L.M. (2004) 'Disaster intervention: long-term psychosocial benefits in Armenia'.*Prehospital Disaster Medicine* 19(1):79–85.

Nakamori, Y., et al. (1997) 'Burn injuries in the 1995 Hanshin-Awaji earthquake.' *Burns*, 23(4): 319–22.

Naoi, M. (1990) 'Subjective well-being of the urban elderly'. Vol.39. *Comprehensive urban studies*.149–159.

Nates, J.L.. Moyer, V.A.. (2005) 'Lessons from Hurricane Katrina, tsunamis, and other disasters', *Lancet*. 366, Oct. 1: 1144–1146.

National Institute of Population and Social Security Research (1998) *The 11th fertility trend basic survey*. Online: http://www.ipss.go.jp/index.html. MOHLW. Accessed October 2002.

Ninomiya, T. (1989) 'Welfare and support for the elderly in the community: from a survey in Sumida-ku, Tokyo, Japan', in: Eekelaar, J.M. and Pearl, D. (eds.) *An aging world: Dilemmas and challenges for law and social policy*. Oxford: Clarendon Press.

Nishikitori, N. Abe, T, Costa, DGM, Dharmaratne, SD, Kunii, O, Moji, K (2006) 'Timing of mortality among internally displaced persons due to the tsunami in Sri Lanka: cross sectional household survey', *British Medical Journa*l, 11 Feb, 332:334–335.

Noji, E., et al. (1993) 'Issues of rescue and medical care following the 1988 Armenian earthquake', *International Journal of Epidemiology*, 22(6): 1070–6.

Noji, E.K (1997a) 'The epidemiology of earthquakes: implications for vulnerability reduction, mitigation and relief', in *Earthquakes and People's Health: Vulnerability Reduction, Preparedness, Rehabilitation*. Kobe: WHO.

Noji, E.K. (1997b) 'The Nature of Disaster: General Characteristics and Public Health Effects', in Noji, E.K. (ed.) *The Public Health Consequences of Disasters*, New York, Oxford: Oxford University Press. 3–20.

Ochiai, E. (1997) *The Japanese family system in transition: A sociological analysis of family change in postwar Japan*. LTCB International Library Foundation: Tokyo.

Oda, J., et al. (1997) 'Analysis of 372 patients with crush syndrome caused by the Hanshin-Awaji Earthquake'. *Journal of Trauma: Injury, Infection, and Criti*cal Care, 42(3): 470–476.

OECD (1996) *Ageing in OECD countries: A critical policy challenge*. Social Policy Studies, OECD.vol.20. Paris.

OECD (2002) *The OECD unemployment outlook*. Paris.

Ogawa, N. (1989) 'Population aging and household structural change in Japan', in: Eekelaar, J.M. and Pearl, D. (eds.) *An aging world: Dilemmas and challenges for law and social policy*. Oxford: Clarendon Press.

Ogawa, N. and Retherford, R.D. (1997) 'Shifting costs of caring for the elderly back to families in Japan: Will it work?' *Population and Development Review*, 23(1): 59–94.

Ogawa, T. (1999) *Decentralisation and diversity in the delivery of social care services for older people in Japan: the development of community care policy*. Sheffield.

Ogino, M. (1999) [Japanese] 'Local municipality's response and local people', in Iwasaki, N. et al. (eds.) [Japanese] *Sociology of the Great Hanshin Awaji Earthquake Vol. 2: Sociology of evacuation life*. Kyoto: Showado.

Okamoto, C. (1994) *Community Welfare Practice Theory: Observation and Perspectives from the Community Welfare Service 20 years*. Tokyo: Chuo Hoki Shuppan.

Onyx, J. and Benton, P. (1995) 'Empowerment and Ageing: Toward Honoured Places for Crones and Sages', in Craig, G. and Mayo, M. (eds.) *Community Empowerment*, London: Zed Books. 46–58.

Orloff, A. S. (1993) *The politics of pensions: A comparative analysis of Britain, Canada, and the United States, 1880–1940*. Madison: University of Wisconsin Press.

Osawa, M. (1995) 'The changing women's employment and the role of social policies in Japan', in B. Koskiaho (ed) *Women, the elderly and social policy in Finland and Japan*, Avebury Ashgate: Aldershot.:127–138.

Otani, J. (1998) 'Intergenerational contracts and rapidly ageing populations in Japan'.in *WHO Symposium on Ageing and Health: A Global Challenge for the 21st Century*. World Health Organisation (WHO): Kobe. Japan.

Otani, J. (2000a), 'Gendered Dynamics of Community Care Development: The Experience of Kobe'. *Journal of Asian Women's Studies*, 9: 14–29.

Otani, J. (2001b), [Japanese] 'The Experience of the ageing society at the post-1995 Great Hanshin Earthquake: Gender perspectives', *Journal of Asian Women's Studies* 10: 32–39.

Otani, J. (2003) 'Older people's experiences after the Kobe Earthquake in Japan: the application of QSR NVivo in non-English context', in *The 2nd Strategies in Qualitative Research (SQR) Conference: Issues and Results from analysis using QSR NVivo and NUD*IST*2000. London Institute of Education.

Otani, J (2006) [Japanese] *Innovative methods of Case Study: Older survivors of the 1995 Great Hanshin Earthquake Five Years On: Implications for a future model of an ageing society with Japanese values*. Fukuoka: Kyushu University Press.

Otani, J (2009a) 'Issues in modern China highlighted by the Sichuan Earthquake in 2008 with reference to the 1995 Great Hanshin Awaji Earthquake and the 2005 West Off Fukuoka Earthquake'. *Bulletin of Kyushu University Asia Center*, 3: 23–37.

Otani, J. (2009b) 'Elderly survivors of the 1995 Great Hanshin Earthquake in Japan' in L. Richards (ed) *Handling Qualitative Data* Second Edition, London. Sage.

Otani, J. (2010) [Chinese] *Life after disasters: Impact of the 1995 Great Hanshin Earthquake on ageing society.* SMC Publishing: Taiwan.

Palmore, E.B. (1999) 'Housing and Health Care', in *'Ageism: Negative and Positive' 2nd edition.* New York: Springer Publishing Company, Inc. 140–151.

Pennec, S. et al (1996) *Women's labour force participation and family size: the case of France and Japan.* Tokyo: Institute of Population Problems, Ministry of Health and Welfare.

Perren, K. Arber, S. and Davidson, K. (2003) 'Men's organisational affiliations in later life: the influence of social class and marital status on informal group membership'. *Ageing and Society*, 23(1): 69–82.

Pharr, S.J. and Krausseds, E.S. (1996). *Media and Politics in Japan.* Honolulu: University of Hawaii Press.

Phifer, J.F. Kaniasty, K.Z. and Norris, F.H. (1988). 'The impact of natural disaster on the health of older adults: a multiwave prospective study'. *Journal of Health and Social Behavior.* 29 (March): 65–78.

Phillips, D.R. (1992) *Ageing in East and Southeast Asia,* London: Arnold.

Potter, J. (1996) 'Discourse analysis and constructionist approaches: theoretical background', in Richardson, J.E. (ed.) *Handbook of qualitative research methods for psychology and the social sciences*, Leicester: BPS Books. 125–140.

Potter, J. and Wetherell, M. (1987) *Discourse and social psychology: Beyond attitudes and behaviour.* Sage.

Powell, M. and Anesaki, M. (1990) *Health Care in Japan.* London and New York: Routledge.

Power, A. (1999) 'High-rise estates in Europe: Is rescue possible?' *Journal of European Social Policy*, 9(2): 139–163.

Pynoos, R.S., et al. (1998), 'A public mental health approach to the postdisaster treatment of children and adolescents'. *Child Adolesc Psychiatr Clin N Am,* 7(1): 195–210.

Pynoos, R.S., et al. (1993), 'Post-traumatic stress reactions in children after the 1988 Armenian earthquake'. *British Journal of Psychiatry*, 163: 239–47.

Quinn, S.C. (2006) 'Hurricane Katrina: A social and public health disaster', *American Journal of Public Health,* 96(2): 204.

Redmond, A.D. (2005) 'Natural disasters: ABC of conflict and disaster', *British Medical Journal,* 330; 28 1259–1261.

Richards, L. (1999a) 'Approaching Qualitative Research', *Week-long workshop: Approaching Qualitative Research"* Australia.

Richards, L. (1999b) *Using NVivo in Qualitative Research.* Melbourne: Qualitative Solutions and Research Pty. Ltd. (QSR).

Richards, L. (1999c) *Qualitative Computing and Qualitative Sociology: the first Decade.* Sociological Research Online.

Richards, L. (2005) Handling Qualitative Data: A practical guide book, London: Sage.

Richards, L. and Richards, T. J. (1991) 'The Transformation of Qualitative Method: Computational Paradigms and Research Processes', in N. Fielding & R. Lee (eds.)*Using Computers in Qualitative Analysis.* Berkeley: Sage.

Richards, T.J. and L. Richards (1994) 'Using computers in qualitative research', in Denzin, N.K. and Lincoln, Y.S. (eds.) *Handbook of Qualitative Research*', Thousand Oaks: Sage. 445–462.

Richardson, L. (1994) 'Writing: A Method of Inquiry', in Denzin, N.K. and Lincoln, Y.S. (eds.) *Handbook of Qualitative Research*', Thousand Oaks: Sage. 516–529.

Ritumeikan University (1998) [Japanese] *Policy study of the earthquake reconstruction.* Yuhikaku.

Robertson, L. and Bell, B. (1976) *Planning for the elderly in natural disaster.* University of Nebraska: Omaha, NE.

Robson, C. (1993) *Real World Research: A Resource for Social Scientists and Practitioner-Researchers.* Oxford: Blackwell.

Robson, C. (2002) *Real World Research: A Resource for Social Scientists and Practitioner-Researchers, Second Edition.* Oxford: Blackwell.

Rosenbaum, S. (2006) 'US Health Policy in the Aftermath of Hurricane Katrina', *Journal of American Medical Association*, 295(4): 437–440.

Rowles, G.D.(1994) 'Evolving Images of Place in Aging and "Aging in Place"', in Shenk, D. and Achenbaum, W.A. (eds.) *Changing Perceptions of Aging and the Aged*', New York: Springer Publishing Company, 115–126.

Rubinstein, E.A. and J.D. Brown (1985) 'The media, social science, and social policy for children'. *Child and family policy*, Norwood, N.J.: Ablex Pub. Corp.

Rubinstein, R. (1986) *Singular paths: Old men living alone.* New York: Columbia University Press.

Rubinstein, R., Kilbride, J.C. Nagy, S. (1992) *Elders living alone: frailty and the perception of choice.* New York: Walter de Gruyter.

Russell, C. and T. Schofield (1999) 'Social isolation in old age: a qualitative exploration of service providers' perceptions'. *Ageing and Society*, 19: 69–91.

Saava, M. and I. Kisper-Hint (2002) 'Nutritional assessment of elderly people in nursing house and at home in Tallinn'. *Journal of Nutrition, Health and Aging*, 6(1): 93–5.

Sato, I. Yoshii, H. Yamada, T. (1998) [Japanese] Fieldnote as methodology: from field enquiry to story making', *Writing Ethnographic Fieldnotes,*University of Chicago Press.

Science News (2001) 'Weak appetite in elderly ties to hormone', in Magazine of the Society for Science and the public. Williamsport, PA.

Life Style Research Institute (1996) [Japanese] *Opinion survey on co-habitation or living in neighborhoods with parents and parents-in-law.* Online: http://www.dai-ichi-life.co.jp/LDI/report/rp9609b.htm#2. Accessed in January 2004.

Shanas, E. (1979a) 'Social myth as a hypothesis: the case of the family relations of old people'. *The Gerontologist*, 19(1): 3–9.

Shanas, E. (1979b) 'The family as a social support system in old age'. *The Gerontologist*, 19(2): 169–74.

Sharan, P., et al. (1996) 'Preliminary report of psychiatric disorders in survivors of a severe earthquake'. *American Journal of Psychiatry*, 153(4): 556–558.

Shimazu, T., et al. (1997) 'Fluid resuscitation and systemic complications in Crush Syndrome: 14 Hanshin-Awaji earthquake patients.' *Journal of Trauma: Injury, Infection, and Critical Care,* 42(4): 641–646.

Citizen and NGO Disaster Prevention International Forum Executive Committee (1998) *The Great Hanshin-Awaji earthquake reconstruction plan made by people: What we can do?*

Silverman, D. (1994) *Interpreting Qualitative Data: Methods for Analysing Talk, Text and Interaction.* London: Sage.

Sonoda, K. (1988) *Health and Illness in Changing Japanese Society.* Tokyo: University of Tokyo Press.

Steinfeld, E. (1981) 'The Place of Old Age: the Meaning of Housing for Old People', in Duncan, J.S. (ed.) *Housing and identity*', Guildford, London, Oxford, Worcester: Billing and Sons Limited. 198–246.

Stevens, N. (2001) 'Combating loneliness: a friendship enrichment programme for older women'. *Ageing and Society*, 21: 183–202.

Stevens, N. and Tillburg, T.V. (2000) 'Stimulating friendship in later life: a strategy for reducing loneliness among older women'. *Educational Gerontology*, 26: 15–35.

Stone, R. Kerr, RA. (2005) 'In the Wake: Looking for keys to Posttraumatic Stress', *Science*, 310: 1605.

Strauss, A.L. and J. Corbin (1990) *Basics of qualitative research: grounded theory procedures and techniques.*. Newbury Park, California: Sage Publications.

Strauss, A. and J. Corbin (1998) *Basics of Qualitative Research: Techniques and Procedures for Developing Grounded Theory. 2nd Edition.* Thousand Oaks: Sage.

Sugama, S. (1999) [Japanese] 'Volunteers at temporary shelters: Case study of Kobe City', in Iwasaki, N. et al. (eds.) [Japanese] *Sociology of the Great Hanshin Awaji Earthquake.* pp.303–321. Kyoto: Showado.

Susser, I. (1993). *The social order of the slum: Ethnicity and territory in the inner city.* Chicago: University of Chicago Press.

Suzuki, K. (2000) *Qualitative Social Research in Japan.* Forum: Qualitative Social Research. 1(1). Online: http://qualitative-research.net/fqs/fqw.htm.

Suzuki, M., B.J. Wilcox, and C.D. Wilcox (2001), 'Implications from and for food cultures for cardiovascular disease: longevity'. *Asia Pacific Journal of Clinical Nutrition,* 10(2): 165–171.

Takahashi, M., and Hashimoto, R. (1997) '*Minsei i'in*—between public and private: a local network for community care in Japan'. *International Social Work*, 40, 303–313.

Takahashi, Y. (1997) 'The self-governing body and community'. *Journal of Urban Studies* 49(11).

Takakura, R., et al. (1997) 'Follow-up after the Hanshin-Awaji Earthquake: Diverse influences on pneumonia, bronchial asthma, peptic ulcer and diabetes mellitus'. *International Medicine*, 36(2): 87–91.

Takayose, S. (1999) *The Great Hanshin Earthquake and Life Reconstruction.* Tokyo: Keiso Shobo.

Tanida, N. (1996) 'What happened to elderly people in the Great Hanshin Earthquake?' *British Medical Journal.* 313(7065): 1133–5.

Taylor, M. (2003) *Public Policy in the Community.* Palgrave MacMillan.

Thomas, A., J. Chataway, and Wuytseds, M. (1998) *Finding out fast: Investigative skills for policy and development.* London: Sage Open University Press.

Ticehurst, S. Webster, R.A. Car, V.J. and Lewin, T.J. (1996) 'Psychological impact of an earthquake on the elderly'. *International Journal of Geriatric Psychiatry.* Nov. 11 (11): 943–951.

Tinker, A. (1997) *Older people in Modern Society, 4th edition.* London and New York: Longman.

Twigg, J. (2000) *Bathing: The body and community care.* London & New York: Routledge.

Ueno, Y. (1997) [Japanese] 'Earthquake-related deaths such as isolated deaths, suicides and overwork deaths', in Kobe University (ed.) *Struggling Evacuation Life*, Kobe: Kobe Shinbun Sogo Shuppan Sentar.

UNDP (2004) *Reducing Disaster Risk: A Challenge for Development.* UNDP.

UN ESCAP (2002) *Population Data Sheet.* Bangkok.

Ungerson, C. (1987) *Policy is Personal: Sex Gender and Informal Care.* London: Tavistock Publications Ltd.

UNICEF (2001) *Buddist Leadership.* UNICEF Thailand newsletter.

UNODC (2000) *Seventh United Nations Survey of Crime Trends and Operations of Criminal Justice Systems, over the period of 1998–2000*, United Nations Office on Drugs and Crime Division for Policy Analysis and Public Affairs. Online: http://www.unodc.org/pdf/crime/seventh_survey/7sc.pdf.

US Centers for Disease Control and Prevention(CDC) (1997) *Principles of Community Engagement.* Atlanta.

Vanholder, R., et al. (2001) 'Intervention of the Renal Disaster Relief Task Force in the 1999 Marmara, Turkey earthquake'. *Kidney International*, 59(2): 783–91.

Vanier, J. (1989) *Community and Growth.* Darton, Longman and Todd Ltd.

Varley, A. (ed.) (1994), *Disasters, Development and Environment.* Chichester: John Wiley & Sons.

Vaughan, R.J. and Buss, T.F. (1998) *Communicating social science research to policymakers.* Thousand Oaks: Sage.

Vaughn, S. (1996) *Focus Group Interviews in Education and Psychology.* Sage Publications.

Walter, T., Littlewood, J. and Pickering, M. (2000) 'Death in the news: the public invigilation of private emotion', in Dickenson, D. (ed.) *Death, dying, and bereavement*', London and Thousand Oaks, CA: The Open University in association with Sage Publications. 14–57.

Warren, C.A.B. and Hackney, J.K. (2000) *Gender Issues in Ethnography 2nd Edition.* Thousand Oaks, London, New Delhi: Sage.

Weintraub, D. and Ruskin, P.E. (1999) 'Posttraumatic stress disorder in the elderly: a review'. *Harvard Review of Psychiatry.* 7(3): 144–152.

Wetherell, M. and Potter, J. (1988) 'Discourse analysis and the identification of interpretative repertoires', in Antaki, C. (ed.) *Analysing everyday explanation: a casebook of methods*, London: Sage.168–183.

Wetherell, M., Taylor, S. and Yates, S.J. (eds.) (2001) *Discourse as Data: A guide for analysis*. Sage in association with the Open University.

Weyerhaeuser (2005) *Hurricane Katrina, Rebuilding a community: an employer's guide to assisting employees*. September.

WHO (1997) 'Earthquakes and People's Health: Vulnerability reduction, preparedness, rehabilitation' in *WHO Symposium on Earthquakes and People's Health*, World Health Organization Centre for Health Development: Kobe Japan.

Wilkinson, R.G. (2005) *The impact of inequality: How to make sick societies healthier*. London: Routledge.

Wilson, G. (2000) *Understanding old age: critical and global perspectives*. London: Sage.

Wolcott, H.F. (1990) *Writing up qualitative research*. 1st Edition. Newbury Park: Sage Publications.

Wolcott, H.F.(2001) *Writing up Qualitative Research*. 2nd Edition. Thousand Oaks, California: Sage.

Woods, P. (1999) *Successful writing for qualitative researchers*. New York: Routledge.

Yamamoto, M. and K. Mizuno (1996) *Influence of Earthquake Damage on Treatment of Patients at Kobe University Hospital*. Online: http://www.kobe-u.ac.jp/~quake/subtheme4/13/13.html]

Yamamoto, N. and M.I. Wallhagen (1997) 'The continuation of family care-giving in Japan'. *Journal of Health and Social* Behaviour, 38:164–176.

Yamamoto, N. and Wallhagen, M.I. (1998) 'Service use by family caregivers in Japan'. *Social Science and Medicine*, 47(5): 677–691.

Yamasaki, S. (1998) 'The challenge of the aging Japanese society: public long-term care (LTC) insurance'. in *Salzburg Seminar: The Challenges for the Ageing Society*. Salzburg, Austria.

Yazawa, S. and Kunihiro, Y. (1999), 'A super-aging regional community and the feminization of politics and social welfare: A case study of Oshima-cho, Oshima-gun, Yamaguchi Prefecture'. *Journal of Asian Women's Studies*, 6 and 7: 65–84.

Yin, R.K. (1994) *Case study research: Design and methods*. Thousand Oaks: Sage Publications.

# Index